MEASURES OF DOUBLE STARS

PUBLICATIONS
OF THE
ASTRONOMICAL OBSERVATORY
UNIVERSITY OF MINNESOTA

VOLUME I

Francis Preserved Leavenworth
1858–1928

MEASURES OF DOUBLE STARS

BY

FRANCIS P. LEAVENWORTH

Late Professor of Astronomy
University of Minnesota

WITH WHICH ARE INCLUDED
THE MEASURES BY
WILLIAM O. BEAL

MINNEAPOLIS
THE UNIVERSITY OF MINNESOTA PRESS
1930

STAFF OF THE OBSERVATORY

CLIFFORD C. CRUMP, Ph.D. *Chairman and Professor of Astronomy*
JOHN S. ALLEN, M.A. *Instructor*
CHRISTINE WESTGATE, B.S. *Assistant*

PRINTED IN THE UNITED STATES OF AMERICA
AT THE LUND PRESS • MINNEAPOLIS • MINNESOTA

INTRODUCTION

This series of measures of double stars has been brought together to form a complete record of the work done by Francis P. Leavenworth in this his chosen field of astronomical research. It covers a period of over forty years and consists of 2,988 annual means of the measures of 1,185 stars. Most of the observations recorded here have been published. They will be found in the *Astronomical Journal;* the publications of the Leander McCormick Observatory, and the Haverford College Studies. This series, however, includes several hundred measures made at the University of Minnesota, which are published here for the first time.

The chief telescope used for this work was the 10½-inch refractor of the University of Minnesota, but many measures were made with other refracting telescopes, the 40- and 12-inch of the Yerkes Observatory, the 10-inch of the Haverford College Observatory, the 26-inch of the Leander McCormick Observatory and the 16-inch of the Goodsell Observatory.

The method of measurement was that followed by many double star observers. Eight settings were made for position angle, the stars being bisected by a single wire or in some instances with the stars midway between two wires. Four measures were made of the double distance. The magnification used was the greatest the conditions would permit, especially when the stars to be measured were exceptionally close.

The position of each star is given for 1880 and, in accordance with the agreement adopted by the International Astronomical Union, for the Standard Equinox of 1950. These positions have been derived primarily from Burnham's General Catalogue or from the discoverer's publications. Some of the positions were kindly furnished by Professor van Biesbroeck of the Yerkes Observatory.

In an appendix the measures of double stars made by Professor William O. Beal have been included. The method of observation was that adopted by Professor Leavenworth. This list contains 341 annual means of the measures of 289 stars. A list of the publications of Francis P. Leavenworth appears at the end of the volume.

The work of collecting the observations was begun under the direction of Dr. Beal by John S. Allen, M. A., who with the writer has carried the work to its completion.

CLIFFORD C. CRUMP
Chairman and Professor of Astronomy

University of Minnesota,
June 3, 1930

v

FRANCIS PRESERVED LEAVENWORTH

Francis P. Leavenworth was born at Mount Vernon, Indiana, September 3, 1858. He was graduated from Indiana University with the degree of Bachelor of Arts in 1880 and some years later, in 1888, he received the Master of Arts degree from the same institution. In the beginning of his astronomical career he was connected successively with the Cincinnati Observatory and the Leander McCormick Observatory of the University of Virginia. He was made director of the Haverford College Observatory in 1887. From this position he came to the University of Minnesota in 1892 as assistant professor of astronomy. In 1897 he was appointed professor of astronomy and he held this position until his retirement as Professor Emeritus in 1927.

Professor Leavenworth was a member of the Royal Astronomical Society, the American Astronomical Society, and the honorary fraternities, Phi Beta Kappa and Sigma Xi.

While Mr. Leavenworth was at the Cincinnati Observatory he had as his fellow students, Herbert C. Wilson, Albert S. Flint, and Herbert A. Howe. Professor Ormand Stone was then the director of the Cincinnati Observatory, and when he left to become the director of the new Leander McCormick Observatory of the University of Virginia, he took Mr. Leavenworth with him as his assistant. At Cincinnati Mr. Leavenworth began the observation of double stars that he carried on throughout the remainder of his life. At the McCormick Observatory he discovered and recorded the positions of 250 new nebulae with the 26-inch Clark telescope. While he was director of the Haverford College Observatory, he made extended observations on four double stars to determine relative parallax. His work was very successful, and the results obtained have an accuracy that is comparable with those obtained by the modern photographic methods.

The 10½-inch objective furnished by the John A. Brashear Company for the telescope at the University of Minnesota can be changed from a visual to a photographic objective by replacing the front lens of the system by a third lens. With this instrument used as a photographic telescope Professor Leavenworth participated in the Eros campaign of 1901 to determine the solar parallax. The advent of Nova Aquila No. 3 caused Professor Leavenworth to turn his attention to irregular variables, and for ten years he divided his observing time between double star and variable star observations. He spent the summer of 1893 at the Goodsell Observatory and the summers of 1914, 1916, and 1919 at the Yerkes Observatory.

His researches during these periods were almost entirely confined to the measurement of double stars.

It was as a teacher of astronomy, however, that Professor Leavenworth performed his most significant work. As such he was never a taskmaster; rather an interpreter of nature, inspiring in others love for the science which meant so much to him. His patience, his faith in human nature, and his unfailing good humor were qualities that won the affection of his students. The very considerable number of young men who, under his influence, devoted their lives to the pursuit of astronomy is striking testimony to his worth and efficiency as a teacher.

The old stone Observatory, built at the period when Professor Leavenworth began his work at the University of Minnesota, still stands upon the campus. To those who recall that here for over thirty-five years he kept his vigils, it serves as a monument to his memory. The structures now crowding around presage that it too will soon be gone. As something more enduring than stone, it seems fitting that in recognition of the value of a sincere life wholeheartedly devoted to science and teaching, the present volume be published by the University he served so long and so well. Had he himself had the choice, he would have chosen such a memorial.

MEASURES OF DOUBLE STARS

MEASURES OF DOUBLE STARS

BGC *	Double Star	R. A. 1880 1950	Dec. 1880 1950	Epoch	Position Angle	Distance	Magnitudes	Nights Observed	Apt.
12755	Σ 3062	23ʰ 59.9ᵐ	57°46′	1907.00	355.7±0.4	1.52±0.02	6.8– 7.5	2	10½
		0 3.5	58 10	1921.01	19.0±1.0	1.44±0.03	7.0– 7.8	3	10½
				1922.02	22.5±0.2	1.50±0.00	6.2– 6.8	2	10½
10	Σ 3063	1.5	— 5 13	1886.46	220.9±0.3	1.97±0.07	8.2– 8.9	3	26
		5.1	— 4 49	1920.87	216.0±1.0	1.61±0.08	8.5–10.2	2	10½
30	β 391	3.2	—28 39	1888.90	273.5±0.3	0.92±0.04	6.0– 6.1	4–3	10
		6.8	15						
32	O. Stone 1	3.5	—14 51	1887.97	107.8	9.94	8.5– 8.7	1	10
		7.1	27	1920.85	108.6±0.6	9.44±0.03	8.5– 8.6	2	10½
41	β 253	4.1	57 51	1916.80	41.0±1.0	0.58±0.05	8.0– 8.3	2	10½
		7.7	58 15	1921.08	33.0	0.73	8.3– 8.3	1	10½
61	β 255	5.6	27 45	1916.41	96.0±4.3	0.53±0.05	7.5– 8.8	4–3	10½
		9.2	28 09	1916.69	90.1	0.43	8.0– 8.7	1	12
				1916.70	89.5	. . . . . .	8.0– 8.7	1–0	40
				1921.83	87.7	0.60	8.0– 9.5	1	10½
70	OΣ 2 AB	7.2	26 19	1914.67	31.8±1.5	0.62±0.04	7.0– 8.5	2	40
		10.8	43	1922.15	19.2±2.0	0.53±0.02	7.0– 8.5	3	10½
				1925.86	24.1	0.61	7.0– 8.0	1	10½
70	OΣ 2 AB & C			1914.67	225.4±0.4	17.98±0.06	. .–10	2	40
81	β 486	8.3	— 8 27	1886.86	3.7±0.3	3.16	6.0–11.5	2–1	26
		11.9	3						
86	β 1027	8.7	20 53	1916.63	187.8±0.2	1.63±0.05	7.5– 9.8	2	40
		0ʰ 12.3ᵐ	21°17′						

* The symbols used in this catalogue are those with which observers of double stars are familiar and will need no explanation here. There is given in order for each star the following data:

(1) Burnham General Catalogue number.
(2) Name of the double star.
(3) Right ascension for 1880 and 1950.
(4) Declination for 1880 and 1950.
(5) Epoch, which is the mean of the times of all the observations made on the particular star in the given year.
(6) Mean of the measures of position angle with the average deviation of all the observations from the mean.
(7) Average distance with the average deviation of all the observations from the mean.
(8) Estimated magnitudes.
(9) Number of nights during the year that the star was observed.
(10) Aperture of the instrument used.

BGC	Double Star	R. A. 1880 1950	Dec. 1880 1950	Epoch	Position Angle	Distance	Magnitudes	Nights Observed	Apt.
126	β 393	0ʰ12.2ᵐ 15.8	−21°48′ 24	1886.81	13.2	. . . . . .	. .−. . .	1–0	26
146	β 256	13.9 17.5	−14 30 7	1886.80 1915.94	248.6±0.2 247.8±1.7	2.55±0.14 2.55±0.08	8.8– 9.1 9.0– 9.2	2 4–3	26 10½
146	β 256 AB & C			1916.00	294.5	65.41	. .–12.0	1	10½
150	β 1015	14.4 18.0	11 39 12 2	1888.56 1896.63 1914.77 1916.63	114.8 127.5 141.9±2.1 139.1±2.1	0.59 0.52 0.54±0.06 0.38±0.01	8.0– 8.0 8.3– 8.6 . .−. . . 8.5– 8.8	1 1 2 3	10 10½ 40 40
151	A. Clark 1	14.6 18.2	32 19 42	1907.85 1914.73 1921.92 1926.78	282.7±0.4 287.5±0.5 285.6±1.1 283.4	1.08±0.02 1.39±0.06 1.31±0.03 1.42	7.5– 8.2 7.5– 8.6 7.9– 8.9 7.5– 8.8	2 4 3 1	10½ 40 10½ 10½
153	β 1093	14.7 18.3	10 19 42	1916.63 1916.69	76.2±3.0 77.9	0.50±0.03 0.46	7.5– 8.7 7.5– 8.7	2 1	40 12
156	β 777	14.9 18.5	− 0 55 32	1914.76 1914.80	166.3 166.9±0.7	3.96 4.15±0.14	9.0– 9.8 8.8–10.8	1 3	40 10½
160	h 1020	15.3 18.9	26 18 41	1914.70	181.5±0.2	11.34±0.20	8.4–11.0	3	40
181	β 488	17.9 21.5	− 4 8 − 3 45	1886.73 1914.58	346.9±1.0 345.2±0.1	3.23±0.12 3.31±0.07	7.6–10.9 7.0–10.2	4 2	26 40
208	β 1225	20.9 24.5	20 26 49	1916.63	190.0±0.7	1.26±0.02	8.2–11.2	2	40
239	β 1095	23.8 27.5	29 5 28	1919.67	358.4±1.3	2.56	6.0–14.0	2–1	40
247	β 107 AB	24.5 28.4	62 41 63 4	1916.08	352.1±1.2	5.91±0.20	9.3–10.4	4	10½
247	β 107 AC			1916.08	336.4±0.4	46.94±0.19	. .– 8.5	4	10½
247	β 107 AD			1916.08	146.2±0.6	50.57±0.22	9.5– 8.7	4–3	10½
247	β 107 AE			1916.08	170.8±0.0	114.08±0.26	9.5– 9.0	4	10½
247	β 107 AF			1916.08	113.5±0.1	152.01±0.22	. .– 9.2	4	10½
247	β 107 AG			1916.86	266.6±0.2	21.15±0.65	. .–12.2	3	10½
247	β 107 AH	0ʰ28.4ᵐ	63° 4′	1916.86	92.9±1.2	45.57±0.51	. .–11.5	5–4	10½

BGC	Double Star	R. A. 1880 1950	Dec. 1880 1950	Epoch	Position Angle	Distance	Magnitudes	Nights Observed	Apt.
260	OΣ 12	0ʰ25.2ᵐ 29.0	53°52′ 54 15	1921.00 1923.98	159.7±0.9 159.8±0.1	0.59±0.00 0.58±0.02	5.5– 5.8 5.7– 5.6	3 2	10½ 10½
276	β 1310 AB	26.4 30.1	22 32 55	1919.66	212.6	3.93	7.0–13.7	1	40
276	β 1310 AB&C			1916.65	297.1±0.2	16.60±0.14	7.0–14.0	2	40
276	β 1310 AC			1919.67	293.4±0.8	16.66±0.18	7.0–13.0	2	40
276	β 1310 AD			1916.65	146.4±0.2	96.01	. .–. . .	2	40
298	Δ 2	28.4 32.0	— 5 12 — 4 49	1886.61	244.6±1.4	0.74±0.12	6.5– 7.0	4	26
314	Ho 212	29.1 32.7	— 4 15 — 3 52	1914.75 1916.62	263.7 286.2	0.28 0.27	6.0– 7.0 . .–. . .	1 1	40 40
314	Ho 212 AB&C			1914.75 1916.62	48.9 46.5	26.46	6.0–13.0 . .–. . .	1 1–0	40 40
332	β 230	31.0 34.7	26 39 27 2	1915.80	323.5±0.7	3.89±0.27	8.6–10.5	3	10½
335	β 395	31.2 34.7	—25 26 3	1886.87 1888.89 1891.76 1914.80	104.2±0.4 109.7±3.3 115.2 109.7±1.3	0.58±0.02 0.66±0.01 0.69 0.78±0.08	6.1– 6.3 6.0– 6.4 . .–. . . 6.5– 6.8	2 3–2 1 3	26 10 10 10½
359	β 257	33.6 37.4	46 36 59	1916.82	243.6±1.9	0.61±0.07	8.0– 8.9	3	10½
364	β 109 AB	34.5 38.0	—17 10 —16 47	1916.32 1917.92	354.5±0.0 354.5	104.64±0.11 104.18±0.30	7.2–10.8 7.0–10.5	2 1–2	10½ 10½
364	β 109 AC			1916.16 1917.92	356.0±0.3 355.8	93.12±0.56 92.74±0.40	. .–11.2 . .–11.0	3 1–2	10½ 10½
364	β 109 BC			1916.09 1917.97	159.7±1.1 158.6	11.71±0.11 11.65	10.7–11.2 10.3–10.8	4 1	10½ 10½
374	OΣ 18	36.2 39.8	3 31 54	1896.81 1914.62	130.1 148.6±0.4	1.07 1.31±0.03	7.5– 9.5 7.5– 9.9	1 3	10½ 40
395	β 231	38.0 41.9	47 38 48 1	1916.82	302.1±0.3	33.23±0.26	5.0–11.1	4	10½
420	β 494	40.9 0ʰ44.5ᵐ	— 1 54 — 1°31′	1886.78 1888.64 1906.97	171.9±0.5 172.1±1.5 171.9±1.1	1.26±0.01 1.29±0.03 1.14±0.02	7.9– 7.9 8.2– 8.3 8.5– 8.8	2 2 2	26 10 10½

BGC	Double Star	R. A. 1880 1950	Dec. 1880 1950	Epoch	Position Angle	Distance	Magnitudes	Nights Observed	Apt.
426	Σ 60	0ʰ41.7ᵐ	57°11′	1907.02	233.6±0.9	5.95±0.14	4.0– 7.0	5	10½
		45.7	34	1920.00	257.0	7.31	. .–. . .	1	10½
				1921.07	258.8±0.2	7.25±0.16	3.8– 7.3	2	10½
				1923.97	261.9±0.2	7.51	4.0– 7.4	3	10½
440	β 232 AB	43.6	49 59	1896.63	331.0	0.47	8.0–. . .	1	10½
		47.5	50 22						
440	β 232 AB & C			1916.80	293.6±0.3	27.48±0.10	8.0– 9.5	2	10½
				1918.00	293.7	27.57	7.7– 9.5	1	10½
456	Σ 67	45.9	9 57	1888.74	4.4	1.71	8.0– 8.4	1	10
		49.5	10 20	1920.87	0.5±0.4	1.98±1.0	8.3– 8.6	3	10½
463	O. Stone 3	46.3	−23 16	1886.87	271.4±0.6	2.14±0.04	7.2– 8.1	2	26
		49.8	−22 53	1920.89	266.6	2.02	7.0– 8.2	1	10½
				1922.83	259.0	1.89	7.0– 8.0	1	10½
479	OΣ 20	48.2	18 32	1886.66	351.5	0.45	7.0– 7.0	1	26
		51.9	55	1891.78	341.8	0.4	. .–. . .	1	10
				1906.86	313.0	0.43	6.0– 7.0	1	10½
				1914.74	307.9±0.5	0.54±0.04	6.1– 6.7	3	40
				1914.89	306.7±4.2	0.48±0.05	6.0– 7.0	3	10½
				1918.05	302.2±0.2	0.56±0.04	6.0– 7.2	2	10½
				1919.07	298.9±1.0	0.51±0.06	6.2– 7.3	3	10½
				1919.64	301.1±0.6	0.55±0.06	6.0– 6.8	4–3	40
				1920.86	299.9±0.6	0.58±0.08	6.0– 6.8	2	10½
				1925.33	290.4±0.6	0.58±0.02	6.0– 7.2	2	10½
. . .	A 2208	48.5	18 49	1914.74	85.1±1.4	1.54±0.04	9.0–10.5	2	40
		52.2	19 12						
. . .	A 2209	48.9	18 17	1914.74	155.2±1.8	1.54±0.04	9.6– 9.9	2	40
		52.6	40						
482	Σ 73	48.5	22 59	1888.50	7.6	1.19	5.7– 6.2	1	10
		52.2	23 22	1906.55	26.2±1.5	0.99±0.06	6.5– 6.4	5	10½
				1914.70	39.7±0.6	1.00±0.03	6.0– 6.5	3	40
				1914.93	38.7±0.8	0.76±0.11	6.3– 6.4	4	10½
				1918.06	44.0±0.8	0.85±0.09	6.0– 6.3	4	10½
				1919.07	47.2±0.8	0.74±0.08	6.2– 6.0	4	10½
				1920.00	51.4	0.64	6.0– 6.2	1	10½
				1920.88	53.7±0.5	0.78±0.09	6.0– 6.2	3	10½
				1921.92	54.7±1.0	0.70±0.04	6.0– 6.3	4–3	10½
				1922.92	55.7±0.1	0.70	6.0– 6.3	3	10½
				1924.81	62.2±1.1	0.68	6.0– 6.1	1	10½
		0ʰ52.2ᵐ	23°22′	1925.86	66.8	0.70	6.0– 6.5	1	10½

BGC	Double Star	R. A. 1880 1950	Dec. 1880 1950	Epoch	Position Angle	Distance	Magnitudes	Nights Observed	Apt.
485	β 500	0ʰ 48.9ᵐ	30° 1′	1896.73	289.8±3.8	0.62±0.02	8.0– 8.0	2	10½
		52.7	24	1924.81	289.8	0.63	8.0– 8.2	1	10½
487	β 233	49.2	−18 6	1886.35	90.3±1.8	1.48	8.4– 9.3	2–1	26
		52.7	−17 43	1916.46	87.9±0.5	1.23±0.12	8.0– 9.4	4–3	10½
490	Howe 1	49.9	−17 1	1886.32	110.2±1.1	1.81±0.18	8.3– 8.5	4	26
		53.4	−16 38	1920.89	124.8	1.83	8.5– 9.0	1	10½
527	β 867	53.9	11 17	1889.00	173.0	0.98	8.0– 9.0	1	10
		57.6	40	1906.91	172.1	0.87	8.5–10.0	1	10½
531	β 234 AB	54.6	−17 43	1887.97	153.0±0.2	4.74±0.23	8.5– 8.5	2	10
		58.1	20	1915.84	150.8±1.6	4.89±0.18	8.2– 8.3	3	10½
531	β 234 AC			1915.84	130.7±0.3	60.91±0.15	. .– 8.8	3–2	10½
551	Ho 213	0 57.4	34 49	1914.75	210.3	0.24	7.5– 7.8	1	40
		1 1.2	35 12						
591	Σ 91	1.0	− 2 22	1886.86	319.8	3.93	7.0– 8.0	1	26
		4.6	− 1 59	1888.41	321.8±0.5	4.05±0.17	6.9– 7.4	2	10
				1920.90	319.8±0.8	3.94±0.07	7.1– 7.9	3	10½
612	β 303	3.2	23 9	1914.92	286.4±0.8	0.64±0.04	7.0– 7.4	2	10½
		7.0	32	1916.95	286.2±1.7	0.65±0.04	7.0– 7.3	2	10½
614	β 235 Aa	3.5	50 22	1916.81	99.1±1.4	0.86±0.03	7.9– 7.5	2	10½
		7.6	45	1921.02	102.5	0.94	. .–. . .	1	10½
				1923.98	102.8±0.5	0.95±0.01	7.0– 7.2	2	10½
				1925.64	103.2	0.95	7.0– 7.1	1	10½
614	β 235 Bb			1916.45	75.9±1.6	8.80±0.23	10.3–12.2	4	10½
614	β 235 Cc			1916.57	47.2±0.9	8.31±0.04	9.7–11.2	3	10½
617	β 2	3.8	29 14	1916.01	153.2±0.8	2.07±0.03	8.8– 9.9	3–2	10½
		7.6	37						
637	β 258 AB	5.6	61 4	1916.73	266.3±4.6	1.04±0.05	6.0– 9.1	6	10½
		10.0	27						
637	β 258 AC			1916.88	203.3±1.8	41.18±0.30	. .–12.2	3	10½
707	Σ 113	13.7	− 1 8	1886.32	350.2±0.8	1.56±0.06	6.0– 6.5	2	26
		17.3	− 0 46	1888.24	349.7±1.3	1.51±0.06	6.3– 6.9	3	10
				1919.88	359.2±0.1	1.36±0.06	6.2– 6.8	2	10½
		1ʰ 17.3ᵐ	− 0°46′	1920.91	359.5±1.0	1.64±0.06	6.2– 6.8	3	10½

BGC	Double Star	R. A. 1880 1950	Dec. 1880 1950	Epoch	Position Angle	Distance	Magnitudes	Nights Observed	Apt.
711	h 2036	$1^h 14.1^m$ 17.6	$-16°26'$ 4	1886.31	22.1 ± 0.4	1.64 ± 0.11	7.0– 7.2	4	26
				1888.75	20.7 ± 1.2	1.71 ± 0.08	7.0– 6.9	3	10
				1906.97	10.7 ± 0.8	1.56 ± 0.16	7.5– 7.6	3	10½
				1915.82	7.2 ± 0.6	1.58 ± 0.13	7.7– 7.8	3	10½
714	β 4 AB	15.0 18.7	10 55 11 17	1886.71	75.2 ± 1.0	0.43 ± 0.02	6.6– 8.3	2	26
				1888.84	61.1 ± 5.9	0.41 ± 0.03	7.1– 7.1	3	10
				1914.75	48.5	0.24	7.0– 7.3	1	40
				1915.80	65.3	0.30	. .–. . .	1	10½
				1919.66	48.4	0.25	. .–. . .	1	40
714	β 4 AB & C			1914.75	249.6	23.03	. .–14.5	1	40
				1919.66	249.9	23.47	. .–13.5	1	40
728	Lv 1	16.8 20.4	1 7 29	1886.69	173.1 ± 0.9	0.80 ± 0.02	9.2– 9.4	2	26
				1914.68	169.0 ± 1.3	0.76 ± 0.02	9.3– 9.6	3–2	40
				1916.62	168.9	0.66	9.4– 9.7	1	40
				1916.69	168.7	0.73	9.4– 9.7	1	12
733	Ho 309 AB	17.6 21.3	19 13 35	1914.72	27.4 ± 0.8	2.90 ± 0.06	8.0–11.8	2	40
733	Ho 309 AC			1914.72	96.0 ± 0.2	44.70 ± 0.12	. .–11.5	2	40
733	Ho 309 AD			1914.72	47.6 ± 0.1	39.60 ± 0.14	. .–14.8	2	40
738	Se 1	17.9 21.2	-24 59 37	1886.89	80.6 ± 0.5	2.59 ± 0.15	7.1– 9.5	3	26
				1888.96	80.9 ± 0.9	3.06	7.5– 9.8	2–1	10
758	β 999 AC	20.5 24.6	44 47 45 9	1916.46	110.5 ± 0.7	122.90 ± 0.42	5.0–11.0	4	10½
758	β 999 CD =β 82			1916.04	140.2 ± 0.3	5.18 ± 0.06	11.0–10.9	3	10½
761	Σ 125	20.8 24.4	$-$ 0 46 24	1891.94	347.5	35.75	. .–. . .		10
765	β 1164	21.4 25.0	4 44 5 6	1914.64	159.4	0.43	7.5– 8.3	1	40
768	β 399	21.8 25.3	-11 31 9	1886.50	301.6 ± 1.5	1.56 ± 0.09	6.0–10.2	3	26
790	β 506	25.1 28.8	14 44 15 6	1914.69	17.4 ± 0.4	1.08 ± 0.04	4.0–11.5	2	40
804	OΣ 31	27.0 $1^h 30.7^m$	7 36 $7°58'$	1914.65	82.3 ± 1.0	3.95 ± 0.08	7.0–10.8	3	40

BGC	Double Star	R. A. 1880 1950	Dec. 1880 1950	Epoch	Position Angle	Distance	Magnitudes	Nights Observed	Apt.
810	Howe 4	1ʰ 27.8ᵐ 31.3	−12°50′ 28	1886.31	145.2±0.4	0.90±0.12	8.3– 8.4	2	26
830	Σ 138	29.8 33.5	7 2 24	1888.48 1920.89	216.2±0.7 222.7±1.1	1.59±0.09 1.56±0.08	7.0– 7.1 7.0– 7.1	3 2	10 10½
836	h 3447	30.6 33.8	−30 31 9	1888.99	89.6±1.1	2.31±0.14	6.0– 7.4	2	10
848	β 508	32.4 36.3	26 20 42	1914.68	63.8±2.2	0.73	9.0–10.5	2–1	40
854	β 5 AB	32.8 38.6	16 1 23	1888.97 1914.73 1915.96 1919.68	292.9 294.0±0.4 290.8±2.9 291.1	1.14 1.19±0.08 0.91 0.91	7.0– 9.0 7.0– 9.2 7.0– 9.2 7.0– 9.0	1 3–2 4–3 2–1	10 40 10½ 40
854	β 5 AB & C			1915.92	185.5±0.4	92.13±0.63	7.5–11.9	5–4	10½
877	Σ 147	35.8 39.3	−11 55 34	1886.75	87.7±0.3	3.57±0.03	6.0– 7.7	2	26
907	Σ 155	37.9 41.6	8 53 9 14	1888.53 1920.93	329.6±0.2 328.0	4.69 4.64±0.03	7.0– 7.2 7.0– 7.2	2–1 2	10 10½
913	β 6	38.7 42.2	− 7 22 1	1886.92 1888.65 1915.82	165.9±0.1 168.6±1.7 165.8±0.4	2.49±0.19 2.72±0.10 2.64±0.04	6.8– 8.8 6.9– 9.6 6.7– 9.1	2 3–2 2	26 10 10½
922	β 784	39.6 43.5	22 18 39	1914.73	46.1	2.17	9.0–10.0	1	40
926	Σ 158	39.8 43.8	32 34 55	1916.70	257.4±1.2	2.14±0.04	8.2– 8.6	2	40
942	β 871	41.8 45.4	− 1 33 12	1886.49 1888.30	352.5±2.0 352.5±1.1	2.21±0.14 1.97±0.01	8.4– 9.2 8.1– 9.4	3 3–2	26 10
956	β 1016	42.9 46.9	32 29 50	1916.69	197.9	0.69	8.5– 9.0	1	40
990	β 259	46.3 49.8	−10 19 − 9 58	1888.97 1920.96	241.2±0.3 238.4±1.0	4.50 4.09	8.0–10.4 8.0–11.2	2–1 2	10 10½
992	β 260	46.8 50.6	14 51 15 12	1916.30 1919.68	242.3±0.5 240.8±1.0	0.82±0.04 0.99±0.03	8.1– 8.7 8.0– 9.0	3 2	10½ 40
995	β 183	47.3 1ʰ 50.7ᵐ	−17 20 −16°59′	1895.95	229.5	2.43	. .−. . .	1	26

BGC	Double Star	R. A. 1880 1950	Dec. 1880 1950	Epoch	Position Angle	Distance	Magnitudes	Nights Observed	Apt.
1002	Σ 183 AB	1ʰ 48.3ᵐ	28°13′	1914.73	354.4±1.2	0.52±0.03	7.5– 8.0	3	40
		52.3	34	1925.86	349.5	0.55	7.5– 7.5	1	10½
1002	Σ 183 AB & C			1914.71	163.9±0.6	5.77±0.10	. .– 9.2	3	40
				1926.02	162.8	4.96	. .–. . .	1	10½
				1927.00	163.6	5.36	. .–. . .	1	10½
1015	Σ 186	49.7	1 15	1886.79	208.0±0.4	0.31±0.00	6.9– 7.1	2	26
		53.3	36	1891.74	230.8	. . . . . .	. .–. . .	1–0	10
				1907.94	32.8	0.71	7.0– 7.0	1	10½
				1914.93	38.0±0.8	0.92±0.07	7.1– 7.0	4–3	10½
				1918.05	39.0±1.0	1.01±0.07	7.2– 7.0	5–4	10½
				1919.07	39.8±0.5	1.00±0.06	7.4– 7.0	4	10½
				1921.03	42.5	1.15	7.0– 7.2	1	10½
				1924.81	42.9	1.24	6.2– 6.0	1	10½
				1925.93	43.8±0.0	1.34	7.1– 7.1	2–1	10½
1034	β 7	51.7	— 2 39	1886.91	11.7±0.6	2.89±0.17	7.0–10.8	3	26
		55.3	18	1915.90	12.8±0.9	2.75±0.15	6.2–11.4	4–2	10½
1039	Σ 194	52.6	24 15	1914.70	271.4±1.2	1.31±0.10	7.9– 8.3	3	40
		56.5	36						
1061	Σ 202	55.8	2 11	1888.26	322.0±0.7	3.00±0.10	3.6– 4.3	5	10
		59.4	31	1906.76	316.9±0.7	2.92±0.1	3.0– 3.7	5–4	10½
				1921.02	311.3±0.6	2.47±0.10	3.0– 4.2	3	10½
				1922.04	309.7±0.1	2.50±0.03	3.0– 4.2	4	10½
				1923.91	310.9	2.24	3.0– 3.8	1	10½
				1924.12	309.4	2.45	3.0– 4.2	2	10½
				1925.66	309.1±1.0	2.32±0.06	3.0– 4.2	3–2	10½
				1926.02	309.2	. . . . . .	3.0– 4.2	1–0	10½
				1927.17	308.4±1.1	2.41±0.14	3.0– 4.0	3	10½
1070	OΣ 38 BC	1 56.5	41 45	1905.97	. . . . . .	0.69	6.0– 9.0	0–1	10½
		2 0.8	42 5	1914.73	111.4	0.55	. .–. . .	1	40
				1916.70	113.1±2.1	0.60±0.04	. .–. . .	2	40
				1921.08	108.9±0.5	0.56±0.02	5.0– 6.5	2	10½
				1927.16	104.6±1.3	0.57±0.01	5.0– 6.0	3	10½
1070	OΣ 38 A & BC			1905.97	62.8	10.51	. .–. . .	1	10½
				1916.70	63.4±0.7	10.14±0.10	. .–. . .	2	40
				1921.08	61.7±0.6	9.84±0.20	4.0–. . .	2	10½
				1922.98	62.1±0.6	9.65±0.04	. .–. . .	3	10½
				1923.56	63.2	9.15	. .–. . .	1	10½
				1927.16	61.2±0.4	9.71±0.09	. .–. . .	3	10½
1091	β 516	1 59.1	— 1 33	1886.84	286.5±2.5	0.78±0.13	8.4– 8.5	3	26
		2ʰ 2.7ᵐ	— 1°13′	1888.85	282.2	0.67±0.07	7.8– 8.5	3	10

BGC	Double Star	R. A. 1880 1950	Dec. 1880 1950	Epoch	Position Angle	Distance	Magnitudes	Nights Observed	Apt.
1098	Ho 312	2ʰ 0.0ᵐ / 3.9	25° 8′ / 28	1914.73	334.4	1.66	6.5–11.5	1	40
1115	Σ 218	2.6 / 6.2	— 1 0 / — 0 40	1888.88 / 1920.96	249.8±0.5 / 248.2±0.2	4.76±0.12 / 4.74±0.15	7.2– 8.0 / 7.0– 8.0	2 / 2	10 / 10½
1139	Σ 226	5.4 / 9.3	23 24 / 44	1914.70	243.7±0.1	2.29±0.10	7.8– 9.5	3	40
1144	Σ 228	6.4 / 10.9	46 55 / 47 15	1921.10	121.8	0.59	7.0– 8.0	1	10½
1179	Hastings	10.0 / 13.3	—18 47 / 27	1886.87 / 1888.50 / 1889.95 / 1891.03 / 1891.76	336.7±0.1 / 340.2±0.4 / 340.5±0.2 / 340.3±0.8 / 341.7±0.8	1.96±0.19 / 2.18±0.05 / 2.36±0.10 / 2.21±0.00 / 2.09	8.0– 8.5 / 8.2– 8.8 / . .–. . . / . .–. . . / . .–. . .	3 / 7 / 4–3 / 3–2 / 2–1	26 / 10 / 10 / 10 / 10
1226	β 8	15.0 / 18.7	8 20 / 39	1888.95 / 1915.82	206.7±1.0 / 207.3±0.2	1.07±0.02 / 1.00±0.11	7.9– 9.1 / 7.9– 9.1	5–3 / 3	10 / 10½
1247	Ho 313 AB	17.4 / 20.9	— 8 23 / 4	1914.72	77.4±0.2	1.94±0.10	8.6– 8.9	2	40
1247	Ho 313 AC			1914.78	86.4	17.46	. .–12.0	1	40
1253	Ho 314	18.1 / 21.6	— 8 25 / 6	1914.78	201.4	4.24	8.8–10.3	1	40
1262	Σ 262 AB	19.2 / 24.9	66 52 / 67 11	1921.15 / 1922.51	249.8±0.8 / 251.1±1.2	2.24±0.11 / 2.19±0.02	4.3– 6.8 / 4.0– 7.0	4 / 2	10½ / 10½
1262	Σ 262 AC			1921.15 / 1922.37	109.9±0.9 / 110.4±1.0	7.17±0.11 / 7.11±0.15	. .– 7.2 / 4.0– 7.5	4 / 3	10½ / 10½
1288	β 519	23.6 / 27.1	— 2 48 / 29	1886.40	57.1±0.4	0.99±0.02	8.3– 9.5	3–2	26
1305	Howe 6	25.8 / 29.3	— 8 5 / — 7 46	1886.84 / 1888.98 / 1914.66	207.3±0.7 / 212.0 / 207.4	2.46±0.03 / . . . . . . / 2.34	9.0– 9.1 / 9.3– 9.6 / 9.5– 9.8	2 / 1–0 / 1	26 / 10 / 40
1308	h 652	26.2 / 29.9	9 3 / 22	1888.01 / 1920.97	316.5 / 314.5±1.2	4.73 / 4.65±0.21	9.3– 9.6 / 8.8– 9.1	1 / 3	10 / 10½
1319	Σ 280	28.1 / 31.6	— 6 10 / — 5 51	1886.74 / 1921.08	346.3±1.4 / 346.6±0.2	3.56±0.04 / 3.30±0.09	7.5– 7.5 / 7.2– 7.2	2 / 2	26 / 10½
1338	β 520	30.8 / 2ʰ 34.3ᵐ	— 4 6 / — 3°47′	1886.94	201.2	0.79	8.6–10.5	1	26

BGC	Double Star	R. A. 1880 1950	Dec. 1880 1950	Epoch	Position Angle	Distance	Magnitudes	Nights Observed	Apt.
1354	β 1315 AB	$2^h\,32.5^m$	13°59′	1914.73	130.0±1.1	1.78±0.02	8.0– 9.8	3	40
		36.3	14 17						
1354	β 1315 AC			1914.75	56.6±0.0	77.93±0.00	. .– 9.4	2	40
1354	β 1315 CD			1914.75	30.2±0.1	6.57±0.20	. .–13.8	2	40
1365	OΣ 43	33.7	26 6	1914.71	40.2±1.0	1.26±0.04	7.0– 9.8	2	40
		37.8	24	1919.08	35.6±0.8	0.98±0.03	7.0– 9.0	4	10½
				1920.98	35.0±0.4	1.06±0.02	7.0– 8.8	2	10½
1386	Σ 295	35.1	— 1 12	1888.51	324.1±1.1	4.28±0.05	6.0– 9.6	2	10
		38.7	— 0 54						
1398	β 306 AB	36.9	25 8	1914.72	21.6±0.2	2.86±0.08	6.4–10.8	2	40
		40.9	26	1916.92	16.4	2.98	6.0–11.3	1	10½
1398	β 306 AC			1914.78	90.4	50.22	6.5–11.0	1	40
1401	Σ 299	37.1	2 44	1888.09	291.0±1.3	3.06±0.14	4.0– 7.5	4–3	10
		40.7	3 2	1920.93	290.9±0.4	2.92±0.08	3.0– 7.2	2	10½
1409	β 261 AB	38.6	—28 25	1916.35	96.8±2.9	3.18±0.02	7.8–10.8	3	10½
		41.6	7						
1409	β 261 AC			1916.35	130.6±0.4	70.29±0.29	. .– 9.6	3	10½
1418	β 9	39.7	35 3	1915.94	171.5±0.8	1.81±0.08	6.3– 8.7	3	10½
		44.0	21	1922.06	173.2±1.4	1.82±0.11	6.0– 8.6	2	10½
1424	β 262	40.6	30 33	1916.10	54.5±1.1	1.26±0.05	7.9–10.3	3	10½
		44.8	51	1916.93	59.2±1.6	1.58±0.12	8.0–10.3	2–3	10½
1420	β 83	40.0	— 5 28	1886.84	116.4±1.2	0.88±0.05	7.5– 8.8	3–2	26
		43.5	10	1888.87	109.2±0.9	0.99±0.01	7.8– 9.8	2	10
				1916.87	99.4	0.81	8.0– 9.5	1	10½
				1918.05	101.0±2.2	0.74±0.06	7.9– 9.4	2	10½
				1919.06	95.7±2.5	0.90±0.02	8.0– 9.8	2	10½
1427	Σ 305	40.7	18 52	1888.92	317.5	. . . . . .	. .–. . .	1–0	10
		44.6	19 10	1914.94	315.8±0.9	3.20±0.12	7.0– 7.5	2	10½
				1918.06	315.6±0.6	3.17±0.07	7.0– 7.5	5	10½
				1919.06	315.1±0.8	3.15±0.02	7.0– 7.8	4	10½
				1920.00	313.7	3.12	7.3– 7.9	1	10½
				1920.99	314.5±0.7	3.10±0.05	7.3– 7.9	3	10½
				1922.04	313.4±0.4	3.13±0.05	7.3– 7.9	3	10½
				1923.98	313.4±0.4	3.20±0.07	7.0– 7.7	2	10½
				1925.13	312.8	3.07	7.0– 8.0	1	10½
		$2^h\,44.6^m$	19°10′	1926.02	312.9	3.58	7.0– 7.6	1	10½

BGC	Double Star	R. A. 1880 1950	Dec. 1880 1950	Epoch	Position Angle	Distance	Magnitudes	Nights Observed	Apt.
1444	Lv 2	$2^h 42.1^m$ 45.3	$-18°49'$ 31	1886.82	25.2 ± 0.8	2.92 ± 0.16	8.2–11.5	2	26
1454	Σ 315	43.5 46.9	-11 3 -10 45	1886.91 1923.91	157.2 ± 0.2 158.8	2.46 ± 0.6 2.33	7.9– 8.7 7.0– 8.5	2 1	26 10½
1472	Σ 323	46.3 50.0	5 59 6 17	1914.73	280.4 ± 1.2	2.74 ± 0.08	7.8– 8.0	2	40
1508	β 525	52.0 56.0	21 8 25	1914.71 1919.66	159.0 ± 6.2 183.0	0.26 ± 0.08 0.26	7.5– 7.8 7.0– 7.0	2 1	40 40
1512	Σ 333	52.4 56.4	20 52 21 9	1888.86 1907.15 1914.70 1915.45 1918.05 1919.07 1919.67 1920.00 1920.97 1922.08 1925.13 1926.13 1927.21	202.4 ± 0.2 202.6 ± 0.2 204.4 ± 1.3 202.2 ± 0.8 202.4 ± 0.6 203.2 ± 0.3 203.6 ± 0.3 204.4 203.7 ± 1.0 203.9 ± 1.2 204.0 204.3 202.9 ± 1.1	1.19 ± 0.07 1.29 ± 0.02 1.53 ± 0.07 1.49 ± 0.05 1.40 ± 0.03 1.41 ± 0.06 1.44 ± 0.05 1.38 1.49 ± 0.11 1.31 ± 0.10 1.52 1.44 1.42 ± 0.04	5.7– 6.2 6.2– 6.6 6.0– 6.3 5.8– 6.4 5.9– 6.3 6.0– 6.4 5.8– 6.1 5.5– 6.0 6.0– 6.8 6.0– 6.6 6.0– 6.4 6.0– 6.2 6.0– 6.4	3 2 3 2 5 4 2 1 3 3 1 1 2	10 10½ 40 10½ 10½ 10½ 40 10½ 10½ 10½ 10½ 10½ 10½
1517	Σ 334	53.0 56.7	6 10 27	1888.99 1921.00	317.9 ± 1.1 316.2 ± 0.1	1.34 1.36 ± 0.06	7.5– 7.8 7.8– 8.0	2–1 2	10 10½
1532	Hu 431	54.2 58.2	21 10 27	1919.67	192.2 ± 0.6	1.03 ± 0.04	9.5– 9.8	2	40
1549	β 11	2 56.8 3 0.2	$-$ 8 9 $-$ 7 52	1888.05 1915.87	84.7 83.9 ± 0.5	. . . . . . 2.41 ± 0.04	6.0–10.3 6.2– 9.2	1–0 2	10 10½
1572	Σ 355	0.9 4.6	7 56 8 13	1888.06	150.6	. . . . . .	8.6– 9.3	1–0	10
1580	β 528	2.4 5.9	$-$ 4 3 $-$ 3 47	1886.84 1889.02	13.8 18.2	0.83	8.2– 8.3 8.6– 8.5	1 1–0	26 10
1581	Σ 357	2.6 5.9	-13 3 -12 47	1888.03	297.4	8.51	8.5– 9.3	1	10
1584	β 1030	3.2 7.2	21 17 33	1919.68	145.8	0.64	. .–. . .	1	40
1623	Σ 367	7.9 $3^h 11.5^m$	0 18 $0°34'$	1888.06 1921.04	231.2 ± 1.0 193.6 ± 1.4	0.81 ± 0.04 0.63 ± 0.04	7.4– 7.6 . .–. . .	2 2	10 10½

BGC	Double Star	R. A. 1880 1950	Dec. 1880 1950	Epoch	Position Angle	Distance	Magnitudes	Nights Observed	Apt.
1623	Σ 367	3ʰ 11.5ᵐ	0°34′	1922.02	186.1	0.69	8.0– 8.6	1	10½
				1923.97	182.8	0.78	8.0– 8.3	1	10½
1640	β 84	10.1	— 6 22	1886.71	28.5±1.3	0.69±0.11	6.2– 8.0	3	26
		13.6	6	1888.94	22.4±0.3	0.66	6.4– 7.7	2–1	10
				1915.98	17.3±1.7	0.70±0.02	6.5– 7.2	3	10½
1650	A. Clark 2	12.2	— 1 22	1886.77	Single with power 2000			1	26
		15.8	6						
1699	β 12	18.8	—14 25	1886.94	276.5±0.5	2.12±0.09	7.0– 8.6	1	26
		22.1	10	1888.00	273.9	2.13	7.0– 9.5	1	10
				1915.95	276.5±0.9	2.13±0.22	7.0– 9.8	3	10½
1720	β 878	21.5	22 23	1914.72	77.3±2.2	1.39	6.0–13.5	3–1	40
		25.6	38	1919.66	75.9	. . . . . .	6.0–13.5	1–0	40
1743	Σ 408	24.7	— 4 41	1886.40	337.3±2.1	1.52±0.12	7.8– 7.8	2	26
		28.2	26	1889.02	337.2	1.35	8.0– 8.0	1	10
1761	Σ 412	27.3	24 4	1919.66	95.0	0.24	. .–. . .	1	40
		31.4	18						
1795	β 308	32.1	— 8 2	1885.96	333.4	1.91	. .–. . .	1	26
		35.5	— 7 48						
1838	Barnard 3 AB	37.1	23 43	1914.74	148.7±0.8	1.68±0.02	9.6–10.0	2	40
		41.2	57						
1838	Barnard 3 AC			1914.74	202.0±0.6	18.30±0.04	. .–15.5	2	40
1851	Lv 3	38.5	—13 48	1886.88	0.3±1.7	1.10±0.02	8.2– 9.7	2	26
		41.8	34	1888.93	359.2±0.4	1.08±0.12	8.3–10.2	2	10
1900	OΣ 65	43.1	25 13	1914.73	207.6±1.5	0.74±0.11	6.2– 6.8	3	40
		47.3	26	1921.05	207.0±0.8	0.59±0.06	6.4– 7.3	3	10½
				1922.11	206.9±0.1	0.58±0.01	6.0– 7.0	2	10½
				1925.61	208.0±2.6	0.52±0.02	. .–. . .	2	10½
				1927.16	209.8±0.4	0.42±0.01	. .–. . .	2	10½
1904	Hu 814	43.5	32 12	1914.72	85.7	1.03	8.0–12.5	1	40
		47.9	25						
. . .	h 1408	44.:	—37 58	1888.64	205.0	7.28	. .–. . .	1	10
		47.:	45						
1944	β 263	48.8	32 50	1916.03	76.5±4.2	0.76±0.06	8.4– 8.6	4	10½
		3ʰ 53.2ᵐ	33° 3′						

BGC	Double Star	R. A. 1880 1950	Dec. 1880 1950	Epoch	Position Angle	Distance	Magnitudes	Nights Observed	Apt.
1953	β 542 AB & C	3^h 50.4^m 53.8	$-$ 7°18′ 5	1886.84	194.0	1.50	8.4– 9.4	1	26
2010	Σ 489	56.5 3 59.9	$-$ 7 20 8	1886.89 1921.05	197.4±0.4 197.0±1.2	3.20±0.10 3.00±0.11	8.3– 8.5 8.4– 8.7	2 3–2	26 10½
2034	Σ 493	4 0.4 4.1	5 22 34	1888.96	93.6±0.0	1.68±0.13	8.5– 8.9	2	10
. . .	Jon 236	3.0 6.6	1 6 17	1917.16 1918.07	218.1±0.4 224.9±2.1	4.18±0.26 3.78±0.19	9.3– 9.7 9.6–10.3	3 3	10½ 10½
2084	β 547	7.4 11.2	8 58 9 9	1889.11 1921.06	351.0 356.5	. . . . . . 0.90	5.0– 8.0 5.0– 8.0	1–0 1	10 10½
2093	OΣ 77	8.3 12.7	31 24 35	1919.68	231.7	0.50	. .–. . .	1	40
2100	β 86	8.6 12.8	23 13 24	1916.03	49.8±1.8	4.28±0.07	8.9– 9.5	3	10½
2102	Σ 516	8.7 12.0	$-$10 33 22	1888.54 1921.14 1922.12	150.0±1.0 147.0±0.3 147.8±0.9	6.52±0.10 6.25±0.12 6.46±0.02	6.0– 8.0 6.2– 8.5 6.0– 8.8	2 3 2	10 10½ 10½
2109	Σ 518 AB			1919.10	105.4±0.3	82.72±0.12	. .–. . .	4	10½
2109	Σ 518 BC	9.9 13.3	$-$ 7 47 36	1886.00 1919.09	111.9±0.5 16.7±0.7	3.14±0.15 3.40±0.09	. .–. . . 9.0–11.2	2 3	26 10½
. . .	Espin 168	12.0 16.6	36 26 37	1918.04	282.3	7.34	8.5–11.5	1	10½
2149	β 87	15.3 19.4	20 32 42	1915.93	168.2±1.0	2.11±0.03	6.2– 8.7	2	10½
2154	OΣ 82	15.9 20.4	14 46 56	1914.77	77.6	0.72	7.5– 9.5	1	40
2157	Σ 536	16.2 19.7	$-$ 4 58 48	1886.00	164.4±1.6	1.96±0.10	. .–. . .	2	26
2161	Σ 535	16.6 20.5	11 6 16	1888.41 1920.83 1922.13 1923.33 1927.17	334.7±0.4 320.0±1.1 321.9±0.3 320.6±1.4 314.7	1.38 1.57±0.15 1.55 1.48±0.13 1.44	7.1– 8.3 6.9– 8.5 6.5– 8.0 7.0– 8.5 7.0– 8.8	3–1 4 2–1 4 1	10 10½ 10½ 10½ 10½
2167	β 402	17.0 4^h 20.5^m	$-$ 1 33 $-$ 1°23′	1892.08 1921.06	73.7 71.9	. . . .	8.0–11.0 . .–. . .	1–0 1–0	10 10½

BGC	Double Star	R. A. 1880 1950	Dec. 1880 1950	Epoch	Position Angle	Distance	Magnitudes	Nights Observed	Apt.
2188	Σ 544	4ʰ 19.0ᵐ	— 9° 1′	1888.04	353.3±0.4	2.87	8.0– 8.6	2–1	10
		22.4	— 8 51	1921.13	352.3	2.56	8.0– 8.7	1	10½
2194	β 403	19.3	— 2 20	1888.10	98.4±1.6	1.96±0.02	7.0– 9.2	2	10
		22.8	10	1907.14	96.0±0.2	2.08±0.0	7.0–10.0	2–1	10½
2213	β 311	21.9	—24 21	1891.72	146.2	0.87	7.3– 7.4	1	10
		24.8	11						
2222	β 184	22.8	—21 46	1889.02	262.5	. . . . . .	7.0–10.0	1–0	10
		25.8	36	1891.72	259.2	1.09	7.4– 7.6	1	10
				1916.38	255.2±1.7	1.17±0.03	6.3– 6.9	3	10½
2230	Σ 554	23.3	15 23	1919.15	29.8±1.2	0.80±0.01	6.5– 8.7	4–3	10½
		27.3	33	1921.15	28.6±0.0	0.93±0.06	6.2– 8.6	3	10½
2266	β 1031 CD	29.0	16 16	1914.74	277.1±1.2	1.51±0.08	11.5–14.8	3	40
		33.0	25						
2286	β 185	31.4	—15 10	1916.12	241.3±0.8	2.46	8.0–11.4	3–1	10½
		34.6	1	1917.15	240.0±1.5	2.64±0.03	8.0–11.3	3–2	10½
2336	Σ 589	38.4	5 4	1888.10	299.0±0.5	4.37±0.07	8.0– 8.3	2	10
		42.1	12	1921.06	294.4±0.4	4.46±0.02	8.1– 8.1	2	10½
2350	β 186	40.2	— 7 12	1885.96	177.4	1.80	. .–. . .	1	26
		43.6	4	1888.04	176.5	1.90	7.0– 9.5	1	10
				1916.05	177.4±2.6	1.36±0.03	7.8–10.8	2	10½
				1917.11	181.5±0.9	1.48	8.0–10.5	3–1	10½
2395	β 1237	46.5	23 21	1914.71	59.0±0.2	4.41±0.16	8.0–11.0	2	40
		50.7	28						
2398	β 316	46.9	— 5 29	1888.55	179.5±0.0	1.16±0.02	7.8– 8.0	2	10
		50.3	22						
2427	β 404	49.8	8 58	1888.69	291.4±2.2	1.50	8.8– 8.9	3–1	10
		53.6	9 5						
2428	OΣ 91	50.0	2 59	1889.06	234.9	0.67	6.0– 6.5	1	10
		53.7	3 6	1921.13	231.9	0.69	7.0– 7.2	1	10½
				1923.15	232.7±2.6	0.62	7.0– 7.2	2–1	10½
2443	Σ 622	51.9	1 29	1888.03	174.4	2.68	8.0– 8.4	1	10
		55.5	36	1921.08	169.6±0.1	2.33±0.02	7.9– 8.0	2	10½
2466	Δ 6	54.2	14 20	1916.10	90.4±0.3	0.93±0.08	8.7– 8.8	3	10½
		4ʰ 58.2ᵐ	14°27′						

BGC	Double Star	R. A. 1880 1950	Dec. 1880 1950	Epoch	Position Angle	Distance	Magnitudes	Nights Observed	Apt.
2494	Σ 636	4ʰ 57.3ᵐ 5 0.6	− 8°50′ 44	1889.09 1923.16	98.1 104.5	. . . . . . 3.69	. .–. . . 7.5– 8.2	1–0 1	10 10½
2523	Hu 446	0.1 4.3	22 34 40	1914.72	182.9±0.3	0.92±0.01	9.0– 9.2	2	40
2535	OΣ 98	1.4 5.2	8 20 26	1888.77 1891.79 1907.19 1920.17 1921.09 1923.64 1925.10 1926.12	194.4±1.0 189.3 164.5±1.4 141.3±1.2 139.4±0.5 137.1±1.8 131.5±2.0 130.8±0.7	1.00±0.07 0.98 0.83±0.07 0.90±0.06 0.99±0.05 0.74 0.98±0.02 1.06±0.04	6.1– 7.3 . .–. . . 6.0– 7.0 6.0– 7.0 6.0– 7.2 6.0– 6.8 6.0– 7.2 6.0– 7.4	4–3 1 2 3 3 1 2 2	10 10 10½ 10½ 10½ 10½ 10½ 10½
2544	β 1047 A&BC	2.2 6.6	27 53 59	1914.72	26.6	. . . . . .	. .–. . .	1–0	40
2544	β 1047 BC			1914.73	86.2±2.6	0.35±0.00	8.0– 9.0	2	40
2548	Σ 634	2.8 11.8	79 5 11	1890.85 1891.30	9.12±0.38 9.20±0.57	16.22±0.01 16.16±0.06	. .–. . . . .–. . .	6–2 6–2	10 10
. . .	Jon 323	3.2 7.1	10 52 58	1921.11	164.7±0.7	3.47±0.14	8.3–10.6	3	10½
2560	Σ 651	4.2 7.6	− 7 13 7	1891.97	49.82±0.52	18.90	8.5– 9.5	2–1	10
2565	β 885	4.9 8.4	− 1 55 49	1888.92	191.4±1.5	0.66±0.01	8.0– 8.6	2	10
2594	Σ 661	7.7 10.9	−13 5 0	1886.87 1923.16	357.8 358.5	2.69 2.74	5.0– 7.5 5.0– 7.5	1 1	26 10½
. . .	Espin 170	8.0 12.6	34 17 22	1917.14	19.1	14.15	8.0–10.5	1	10½
2622	β 318	10.2 13.7	− 3 37 32	1886.11	229.8	0.59	. .–. . .	1	26
2639	β 188 AB	11.8 15.2	− 6 58 53	1916.36	250.2±0.6	34.74±0.20	4.0–10.9	4	10½
2639	β 188 AD			1916.36	61.8±0.9	35.99±0.33	4.0–10.2	4	10½
2639	β 188 BC			1916.12 1917.19	60.9±0.9 43.0	4.11±0.30	. .–12.0 11.0–12.3	4–3 1–0	10½ 10½
2673	β 190 AB	14.6 5ʰ 18.0ᵐ	− 8 9 − 8° 4′	1917.18	354.7±1.4	0.56±0.06	8.2– 8.7	3	10½

BGC	Double Star	R. A. 1880 1950	Dec. 1880 1950	Epoch	Position Angle	Distance	Magnitudes	Nights Observed	Apt.
2690	β 888 AB	5ʰ 16.5ᵐ 21.2	37°16′ 20	1914.72	166.8±1.0	8.46±0.24	. .−. . .	2	40
2690	β 888 AC			1914.72	331.0±0.4	27.35	. .−. . .	2–1	40
2690	β 888 CD			1914.72	342.0±0.8	6.80±0.02	14.2–16.0	2	40
2706	Wn 2	17.7 21.2	− 0 59 55	1885.96 1920.76	170.4 165.2±0.4	2.00 2.24±0.15	. .−. . . 6.2– 6.4	1 3	26 10½
2707	O. Stone 10	17.9 21.2	−10 32 28	1888.55 1923.06	119.4±0.1 120.8	1.08±0.04 1.16	8.1– 8.3 . .−. . .	2 1	10 10½
2712	Da 5	18.4 22.3	− 2 30 26	1888.56 1920.18 1922.06 1923.16 1924.08 1925.16 1926.13 1927.19	84.9±0.2 80.1±0.8 80.4±1.6 79.2±0.8 79.3 83.0 80.1 78.6±1.3	1.14±0.10 1.32±0.04 1.41±0.02 1.45±0.09 1.44 1.35 1.51 1.36±0.06	4.0– 5.5 4.0– 5.4 4.0– 5.2 4.0– 5.4 4.0– 5.0 4.0– 5.2 4.0– 6.0 4.0– 5.6	2 4 3 3 1 1 1 4	10 10½ 10½ 10½ 10½ 10½ 10½ 10½
2769	β 320	23.1 26.1	−20 51 47	1889.07	292.7	2.82	3.0– 9.0	1	10
2808	Σ 734	27.0 30.5	− 1 48 44	1888.04	357.6	2.28	7.0–10.0	1	10
2823	β 13	28.6 32.1	− 4 34 31	1917.18	129.9±5.9	1.23±0.00	8.0–10.7	3	10½
2845	Σ 749	29.6 34.0	26 51 54	1919.19 1920.18 1921.09 1922.15 1923.16 1924.08 1925.15 1926.13 1927.20	346.8±1.4 347.8±0.9 345.6±0.5 343.9±0.3 343.5±1.6 343.6 342.1±1.9 346.0 343.9±1.4	0.91±0.04 0.96±0.01 1.03±0.03 0.98±0.04 0.93±0.01 0.90 1.05±0.22 1.00 0.90±0.06	7.2– 7.0 7.1– 7.0 7.0– 7.0 6.8– 6.7 . .−. . . . .−. . . 7.2– 7.0 7.0– 6.5 7.0– 7.1	5 4–3 3 2 2 1 2 1 3	10½ 10½ 10½ 10½ 10½ 10½ 10½ 10½ 10½
2863	β 89	31.5 35.0	− 1 30 27	1888.47 1917.17	0.7±0.9 1.4±1.0	0.89±0.06 0.68±0.03	8.0– 9.3 8.2– 9.0	2 2	10 10½
2889	β 321 AB	34.0 37.1	−17 55 52	1888.95	141.8±0.6	0.73±0.06	7.1– 8.4	2	10
2889	β 321 CD	5ʰ 37.1ᵐ	−17°52′	1888.95	359.6±0.5	1.40±0.02	8.5– 9.0	2	10

BGC	Double Star	R. A. 1880 1950	Dec. 1880 1950	Epoch	Position Angle	Distance	Magnitudes	Nights Observed	Apt.
2902	Σ 774	5h 34.7m	— 2° 0'	1907.16	156.4±0.7	2.65±0.15	. .—. . .	4–3	10½
		38.2	— 1 57	1920.19	157.2±0.5	2.53±0.08	2.0– 4.5	4–3	10½
				1921.45	157.3±0.3	2.50±0.11	2.1– 5.1	4	10½
				1923.62	157.6±1.8	2.58±0.23	. .—. . .	2	10½
				1925.19	156.7±0.7	2.55±0.05	2.0– 5.0	2	10½
				1926.22	157.4	2.38	. .—. . .	1	10½
				1927.20	156.4±0.2	2.64±0.23	2.0– 5.0	2	10½
2902	Σ 774 AC			1907.18	8.7	58.19	. .– 9	1	10½
2905	β 14	34.8	29 47	1916.12	192.2±1.0	6.12±0.26	7.1–10.5	4–5	10½
		39.3	50						
2961	β 91	40.5	20 54	1916.65	83.4±0.6	1.71±0.22	7.5–10.5	2	10½
		44.7	56						
2970	β 92	41.0	21 4	1916.09	169.2±1.0	9.42±0.04	9.0–10.8	3	10½
		45.2	6	1919.16	171.5±0.8	9.26±0.40	9.0–11.0	3	10½
2979	β 93 AB	41.7	20 59	1916.17	121.9±0.5	60.29±0.09	8.4– 9.4	2	10½
		45.9	21 1						
2979	β 93 BC			1916.16	169.2±2.2	6.01±0.32	9.3–11.1	3	10½
2979	β 93 BD			1916.16	322.7±0.7	9.78±0.17	. .–11.3	3	10½
2980	β 15	41.8	— 2 20	1917.18	178.3±1.2	2.16±0.19	8.0–11.8	4	10½
		45.3	18						
3008	β 94	44.2	—14 31	1888.08	179.5±0.1	2.83±0.06	6.0– 8.4	3	10
		47.4	29	1921.15	173.4±0.6	2.62±0.20	5.8– 8.8	2	10½
				1923.06	174.4	2.90	6.0– 8.5	1	10½
3029	β 95	46.2	— 7 20	1916.17	296.8±0.6	14.23±0.27	8.0–11.8	2	10½
		49.6	19	1917.14	296.8±1.6	14.71±0.07	7.8–11.9	2	10½
3053	S 503 AB	49.2	13 56	1888.83	10.0±1.6	3.17±0.05	. .—. . .	30–25	10
		53.2	57	1889.45	4.2±2.4	3.39±0.13	. .—. . .	37–30	10
				1890.16	359.1±1.1	3.74±0.04	. .—. . .	20–8	10
3053	S 503 AC			1920.22	325.0±0.3	22.40±0.08	6.0– 8.0	3	10½
				1921.19	324.4	23.18	6.0– 8.0	1	10½
. . .	Doo 52	50.9	27 21	1921.12	318.2	. . . . . .	9.0–11.5	1–0	10½
		55.3	22						
3112	h 3823	55.8	—31 3	1888.14	115.0±0.8	3.02±0.05	8.6– 8.6	2	10
		5h 58.4m	—31° 3'						

BGC	Double Star	R. A. 1880 / 1950	Dec. 1880 / 1950	Epoch	Position Angle	Distance	Magnitudes	Nights Observed	Apt.
3116	β 16	5ʰ 56.2ᵐ	−10°36′	1889.03	355.0±0.3	1.96±0.10	5.0– 8.6	2	10
		5 59.5	36	1917.17	355.2±1.0	1.94±0.12	6.0–10.2	2	10½
3186	β 17 AB	6 2.8	−11 8	1889.14	180.2±0.1	3.02±0.04	6.6–10.4	2	10
		6.1	8	1917.15	180.7±0.7	3.18±0.20	6.2–10.8	3	10½
3186	β 17 AC			1917.15	252.7±0.8	9.59±0.16	6.2–11.1	3	10½
3251	β 323	8.7	− 1 41	1889.09	96.8±0.3	2.22±0.10	8.0– 9.7	2	10
		12.2	42						
3256	β 193 AB	9.2	4 0	1917.16	91.6±2.0	18.78±0.18	7.8–11.8	2	10½
		12.9	3 59						
3256	β 193 AC			1917.16	230.1±0.9	58.00±0.02	7.8– 9.5	2	10½
3275	β 18	11.1	−12 0	1886.53	275.7±0.5	1.66±0.08	. .–. . .	2	26
		14.4	1	1916.82	277.8±1.0	1.49±0.03	7.1– 8.8	3	10½
. . .	Rumker 4	12.7	29 42	1921.10	271.5±1.2	6.20±0.16	8.9– 9.4	3	10½
		17.2	41						
3355	β 97	18.5	− 1 21	1889.14	261.8±1.0	0.91	7.2– 8.9	2–1	10
		22.0	23	1907.18	262.8	0.99	7.0– 8.5	1	10½
				1917.16	265.5±0.4	1.00±0.02	7.5– 9.0	2	10½
3357	β 568	18.6	−19 43	1889.07	149.2±3.9	0.79	7.0– 7.8	4–1	10
		21.6	45						
. . .	Jon 595	19.4	11 31	1920.26	39.0±0.2	4.70±0.28	9.0– 9.2	2	10½
		23.3	29						
3420	β 1021	24.1	28 28	1892.17	86.0±1.0	0.68±0.00	8.1– 9.4	2	10
		28.6	26	1926.20	83.5	0.86	8.0– 9.0	1	10½
3452	β 98	26.8	− 5 15	1917.18	145.0±1.4	0.83±0.01	8.0– 8.2	2	10½
		30.2	18						
3460	Σ 932	27.5	14 51	1889.08	330.4	2.12	8.1– 8.0	1	10
		31.5	48	1921.09	324.4±0.2	1.93±0.09	8.0– 8.1	2	10½
				1924.77	323.4±0.7	1.92±0.12	. .–. . .	3	10½
				1926.17	323.7±0.4	1.90±0.13	8.0– 8.3	2	10½
				1927.16	321.3±0.1	1.94±0.00	8.0– 8.2	2	10½
3467	β 194	28.1	38 5	1916.26	279.0±0.0	1.02±0.00	8.1– 8.6	2	10½
		32.9	2						
3493	β 754	30.4	−33 55	1892.15	22.8	. . . . . .	. .–. . .	1–0	10
		6ʰ 32.9ᵐ	−33°58′						

BGC	Double Star	R. A. 1880 1950	Dec. 1880 1950	Epoch	Position Angle	Distance	Magnitudes	Nights Observed	Apt.
3518	OΣ 152	6^h 32.0^m 36.4	28°22′ 19	1892.15	35.2±0.7	0.88±0.01	6.8– 8.2	2	10
				1921.21	32.8	0.86	7.0– 8.5	1	10½
				1923.14	31.4	0.86	7.0– 8.3	1	10½
				1925.19	34.5	1.08	. .–. . .	1	10½
				1926.21	34.6±0.6	0.88±0.00	7.0– 8.5	2	10½
				1927.16	30.4±0.2	0.88±0.02	7.0– 8.2	2	10½
3541	Σ 946	34.2 40.4	59 34 31	1926.31	131.0	4.01	7.0–10.0	1	10½
3557	Σ 955 AB	35.4 38.8	– 7 53 57	1886.11	276.7	1.00	. .–. . .	1	26
				1921.07	275.0	1.00	8.5– 9.5	1	10½
3559	Σ 948 AB	35.6 41.8	59 34 30	1891.92	122.6	1.56	. .–. . .	1	10
				1921.18	108.9±0.2	1.56±0.04	5.0– 5.6	2	10½
				1922.36	108.3±1.0	1.77±0.12	5.0– 5.6	3	10½
				1925.24	105.2±0.7	1.65±0.05	5.0– 5.6	4	10½
				1926.31	106.9	1.65	5.0– 5.8	1	10½
3559	Σ 948 AC			1891.92	306.4	8.64	. .–. . .	1	10
				1921.18	306.2±0.7	8.62±0.01	. .– 6.8	2	10½
				1922.36	306.8±2.5	8.57±0.05	. .–. . .	3	10½
3559	AB & C			1925.24	304.4±0.6	9.30±0.22	. .–. . .	2	10½
				1926.31	306.1	8.37	. .– 7.0	1	10½
3562	OΣ 154	35.9 40.8	40 45 41	1891.76	126.2	26.53	. .–. . .	4	10
				1892.32	125.9	26.54	. .–. . .	4	10
3569	β 19	36.6 39.7	–15 53 57	1888.12	167.2±0.5	3.68±0.08	7.8– 9.4	3	10
				1916.18	165.3±0.6	3.62±0.25	7.1– 9.9	4	10½
3579	β 195 AB	37.4 40.3	–23 7 11	1892.16	214.5±0.3	5.73±0.09	7.2–11.2	3	10
				1917.19	214.9±1.2	6.12±0.13	7.6–11.1	6	10½
3579	β 195 AC			1892.15	178.4	35.04	7.0–12.0	1	10
				1917.19	177.9±0.5	35.83±0.01	. .–12.2	3–2	10½
3596	A. Clark 1 [Sirius]	39.9 43.0	–16 33 37	1886.12	34.2±1.0	. . . . . .	. .–. . .	2–1	26
				1917.19	73.2±1.0	11.06±0.09	. .– 7.8	3–4	10½
				1919.16	69.6±0.9	11.11±0.08	. .– 7.4	7	10½
				1920.18	66.9±0.7	11.34±0.16	. .– 7.4	4	10½
				1921.09	66.4±0.8	11.30±0.08	. .– 7.3	6	10½
				1922.24	61.0±0.9	11.22±0.07	. .–. . .	4–3	10½
				1923.15	61.6±1.1	11.38±0.06	. .–. . .	2	10½
				1924.13	58.8±0.1	. . . . . .	. .–. . .	2–0	10½
				1925.16	57.3±1.2	11.18±0.18	. .–. . .	10	10½
				1926.20	54.8±1.4	11.33±0.24	. .–. . .	7–5	10½
		6^h 43.0^m	–16°37′	1927.16	52.9±1.3	. . . . . .	. .–. . .	2–0	10½

BGC	Double Star	R. A. 1880 1950	Dec. 1880 1950	Epoch	Position Angle	Distance	Magnitudes	Nights Observed	Apt.
3635	β 20	6ʰ 43.4ᵐ	—16° 4′	1886.14	32.2	3.70	8.0–11.0	1	26
		46.5	8	1888.47	35.8±0.4	3.11	8.0–10.6	3–1	10
				1917.18	26.0±1.8	3.01±0.27	8.0–11.0	4–3	10½
3652	β 324	44.8	—23 56	1889.01	206.1	1.82	7.0– 8.5	1	10
		47.7	—24 1						
3659	β 898 AB	45.0	—15 53	1886.11	350.9	3.24	. .–. . .	1	26
		48.1	58						
3701	Σ 984	48.4	32 36	1913.29	161.5±0.5	4.95±0.05	. .–. . .	3	10½
		53.0	31	1925.26	156.4±0.2	4.78±0.12	8.0–10.5	2	10½
				1926.22	161.4	4.91	. .–. . .	1	10½
3715	β 326	50.0	2 28	1888.78	66.3±1.5	1.11±0.08	8.0– 9.2	4	10
		53.6	23						
3746	β 327	52.5	— 2 52	1888.10	95.8±1.5	0.73±0.02	8.0– 8.2	2	10
		56.0	57						
3760	Σ 1007 AB	53.9	12 53	1916.19	207.6±0.1	68.30±0.06	8.4– 8.0	3	10½
		57.8	48						
3760	Σ 1007 BC			1916.20	299.3±0.5	15.45±0.31	8.0–12.0	4	10½
3760	Σ 1007 BD			1916.19	246.0±0.3	22.19±0.26	. .–10.6	3	10½
3764	β 100	54.2	12 34	1888.64	259.6±0.7	3.14±0.07	8.0–10.4	2	10
		6 58.4	29	1916.21	261.8±1.0	3.00±0.16	7.5–11.5	2	10½
3876	Σ 1037	7 5.4	27 26	1919.18	242.5	0.38	. .–. . .	1	10½
		9.7	19	1926.20	354.8	0.49	. .–. . .	1	10½
3892	β 196	6.4	— 5 14	1916.17	189.6±0.2	3.06±0.26	9.2–10.6	2	10½
		9.8	21						
3894	Σ 1043	6.5	— 0 29	1888.05	66.6	2.49	. .–. . .	1	10
		10.1	36	1921.15	248.3	2.16	8.8– 8.7	1	10½
3902	β 197 AB	7.0	— 6 57	1888.10	148.8±1.0	2.14±0.10	8.1– 9.4	3	10
		10.4	— 7 4	1916.22	147.0±0.2	2.00±0.12	8.0–10.0	2	10½
				1917.17	147.8±0.8	2.27±0.12	8.0– 9.8	3–2	10½
3902	β 197 AC			1917.18	19.8±1.0	20.09±0.03	. .–11.5	4–3	10½
3902	β 197 AD			1917.18	184.8±0.2	39.28±0.38	. .–10.8	3–4	10½
3934	β 575	9.4	—15 16	1889.16	200.8	0.71	8.0– 8.0	1	10
		7ʰ 12.6ᵐ	—15°23′						

BGC	Double Star	R. A. 1880 1950	Dec. 1880 1950	Epoch	Position Angle	Distance	Magnitudes	Nights Observed	Apt.
3935	Σ 1056	7ʰ 9.5ᵐ	— 1°39′	1886.14	300.4	3.89	. .–. . .	1	26
		13.0	46	1920.17	298.1±0.4	3.69±0.10	7.9– 9.1	3	10½
3938	Howe 17	9.6	— 0 25	1886.14	315.2	2.78	8.5– 8.5	1	26
		13.2	32	1888.14	315.0±1.4	2.43±0.13	8.6– 8.8	3	10
				1892.10	314.5	2.58	9.0– 9.0	1	10
				1907.18	314.1±1.2	2.50±0.00	8.9– 9.1	3–2	10½
3949	OΣ 170	11.1	9 31	1886.16	110.9±2.4	1.34	7.3– 7.6	3–1	10
		14.9	24	1907.16	105.7±1.2	1.48±0.08	7.5– 7.8	3	10½
				1920.17	104.5±1.0	1.44±0.07	7.5– 7.7	4	10½
				1923.18	102.0±1.1	1.60±0.05	7.5– 7.7	4	10½
				1925.13	102.2±0.9	1.54±0.04	7.2– 7.9	2	10½
				1926.17	103.1±0.7	1.52±0.02	7.5– 7.7	2	10½
3975	β 330	13.4	— 0 41	1888.58	217.3±0.8	1.20±0.06	8.0– 9.5	2	10
		17.0	48	1907.16	214.1±1.7	1.11±0.04	8.2– 9.4	2	10½
3990	Σ 1074	14.4	0 38	1888.57	138.5±0.9	0.79±0.06	7.9– 8.2	2	10
		18.0	31						
3998	β 331	15.0	—24 12	1889.15	113.6±1.5	1.92	8.2– 8.9	2–1	10
		17.9	20						
4008	Lv 4	16.:	—19 30	1889.11	129.2	1.95	9.0– 9.4	1	10
		19.:	38	1920.18	130.6±1.3	1.84	9.0– 9.0	2–1	10½
				1921.10	132.2	1.78	9.0– 9.2	1	10½
4053	β 199	20.0	—20 56	1889.10	21.7±0.5	1.76±0.02	7.8– 8.3	2	10
		23.0	—21 4	1892.10	23.2	1.63	. .–. . .	1	10
				1916.19	19.5±1.1	1.50±0.13	6.9– 8.6	2	10½
				1917.20	20.6±0.4	1.51±0.15	7.2– 7.9	3	10½
4060	β 198	20.6	—20 43	1916.23	211.8	6.30	8.0–11.0	1	10½
		23.6	51	1917.19	215.0±0.4	6.15	8.1–11.2	2–1	10½
4074	β 21	21.6	7 11	1916.21	22.4	3.44	6.0–11.5	1	10½
		25.4	3	1917.17	25.2±0.2	4.17±0.28	6.0–11.2	3	10½
. . .	AG . . .	23.1	26 1	1917.27	170.6±1.2	1.80±0.05	8.9– 9.0	3	10½
		27.4	25 53						
. . .	AG . . .	23.6	25 24	1917.27	297.4±0.9	3.88±0.15	8.8– 9.4	3	10½
		27.9	16						
4108	β 22	25.5	33 7	1916.23	150.2±0.0	6.31±0.03	8.0–11.0	2	10½
		7ʰ 30.0ᵐ	32°59′						

BGC	Double Star	R. A. 1880 1950	Dec. 1880 1950	Epoch	Position Angle	Distance	Magnitudes	Nights Observed	Apt.
4122	Σ 1110 AB	7ʰ27.0ᵐ	32° 9′	1888.74	229.6±0.6	5.89	2.0– 2.7	2–1	10
		31.5	0	1892.38	228.9	5.41	. .–. . .	1	10
				1907.22	220.9±0.6	5.53±0.08	2.0– 3.2	5	10½
				1913.34	220.4±0.3	5.53±0.12	. .–. . .	3	10½
				1918.33	216.3±0.5	5.07±0.12	3.0– 4.3	4	10½
				1919.35	215.9±0.0	4.89±0.12	3.0– 4.3	2	10½
				1920.26	216.0±0.7	4.95±0.02	3.0– 3.9	5	10½
				1921.29	215.7±0.2	4.93±0.06	3.0– 4.2	4	10½
				1922.29	215.1±0.2	4.86±0.13	3.0– 4.8	4	10½
				1923.30	214.6±0.8	4.71±0.10	3.0– 4.5	3	10½
				1925.21	213.1±1.0	4.56±0.09	3.0– 4.3	5	10½
				1926.24	212.4±0.4	4.58±0.04	3.0– 4.4	6	10½
				1927.33	211.6±0.5	4.44±0.11	. .–. . .	2	10½
4122	Σ 1110 AC			1913.34	164.4	73.48	. .–. . .	1	10½
				1918.32	164.9±0.1	73.40±0.08	. .– 8.5	2	10½
				1921.28	164.1±0.4	73.20±0.03	. .– 8.6	2	10½
4164	β 200 CD	30.7	35 19	1916.28	240.2±0.8	1.55±0.07	10.2–11.2	4–2	10½
		35.3	10						
4164	β 200 CE			1916.32	204.8	18.12	10.5–13.0	1	10½
4164	β 200 AB			1916.27	190.8±0.2	100.99±0.17	5.2–10.5	2	10½
4192	β 201	33.7	−20 0	1886.20	331.4	2.83	. .–. . .	1	26
		36.8	9	1888.97	330.9	2.80	7.8– 8.0	1	10
				1921.14	331.2±0.6	2.80	8.0– 8.4	2–1	10½
4193	Σ 1126	33.7	5 30	1889.05	143.6±0.5	1.40±0.05	6.9– 7.4	2	10
		37.4	21	1920.28	149.5±0.5	1.11±0.03	7.0– 7.4	3	10½
				1922.23	150.6±0.4	1.18±0.12	7.0– 7.6	4	10½
				1925.24	150.2±1.0	1.18±0.06	7.0– 7.3	3	10½
				1926.26	151.0±1.9	1.18±0.04	7.0– 7.5	3	10½
				1927.23	151.5	0.99	7.0– 7.5	1	10½
4269	Σ 1146	42.3	−11 54	1886.19	16.1	3.50	. .–. . .	1	26
		45.6	−12 4	1921.18	9.8±0.4	2.88±0.00	5.2– 7.2	2	10½
				1922.25	9.7±0.5	2.84±0.00	5.0– 7.2	2	10½
				1926.26	10.2±0.0	2.90±0.04	5.0– 7.5	2	10½
4279	Ho 37	43.9	− 1 58	1888.64	175.8±0.0	1.53	8.5– 8.6	2–1	10
		47.4	− 2 8	1921.18	174.9±1.8	1.48±0.15	8.5– 8.9	2	10½
4310	β 101	46.2	−13 35	1918.29	286.9	0.60	6.0– 7.5	1	10½
		49.4	45	1920.18	291.2±1.4	0.60±0.01	6.0– 7.2	2	10½
4312	OΣ 182	46.4	3 42	1888.12	35.9±0.7	1.15±0.06	8.1– 8.0	3	10
		7ʰ50.1ᵐ	3°32′	1922.18	29.5±0.5	1.00±0.05	. .–. . .	2	10½

BGC	Double Star	R. A. 1880 1950	Dec. 1880 1950	Epoch	Position Angle	Distance	Magnitudes	Nights Observed	Apt.
4312	OΣ 182	7ʰ 50.1ᵐ	3°32′	1925.13	26.3±0.5	1.05±0.05	7.8– 7.8	2	10½
				1926.28	28.5±1.1	0.98±0.04	8.0– 7.9	3	10½
4364	β 902	52.4	−10 34	1892.20	243.7±2.2	1.17	8.0–11.3	3–1	10
		56.2	45						
4402	Σ 1175	56.1	4 29	1888.18	223.1±0.7	1.98	7.9– 9.1	3–1	10
		59.8	18	1920.20	234.6±0.5	1.46±0.09	8.0– 9.4	4	10½
				1921.12	234.8±0.5	1.44±0.02	8.0– 9.4	2	10½
				1924.92	236.0±1.1	1.50±0.09	7.8– 9.4	4	10½
				1926.00	238.2±0.4	1.51±0.09	8.0– 9.7	4	10½
4405	β 23	56.2	3 26	1917.22	182.7±0.4	2.40±0.14	8.0–11.7	2	10½
		59.9	15						
4409	β 202 AB	57.0	−26 53	1892.18	160.6±0.6	7.60±0.02	7.2–10.2	2	10
		59.9	−27 4	1916.27	160.1	7.73	7.5–10.5	1	10½
				1917.20	161.4±0.2	7.79±0.08	8.0–10.2	3	10½
4409	β 202 AD			1916.74	240.3±0.9	29.86±0.02	. .–11.6	2	10½
4414	β 581 AB	7 57.7	12 38	1920.27	164.5	0.48	. .–. . .	1	10½
		8 1.6	27	1926.28	204.3±6.9	0.45±0.01	8.0– 8.3	3	10½
4414	β 581 AB & C			1920.27	200.0	4.66	. .–11.0	1	10½
				1926.27	198.2±0.4	4.57±0.27	. .–11.2	2	10½
4420	β 903	7 58.2	− 1 31	1886.14	29.9	. . . . . .	8.0– 9.0	1–0	26
		8 1.7	42	1888.14	33.8±0.0	1.68	8.2– 9.6	2–1	10
				1920.18	33.6±1.9	1.56±0.10	8.0– 9.9	3	10½
4421	Σ 1177	7 58.3	27 52	1888.12	350.4	3.63	6.8– 7.3	1	10
		8 2.6	43	1920.28	350.2±0.6	3.41±0.01	7.0– 7.6	2	10½
4459	β 583	3.3	− 6 21	1888.13	70.4±0.8	1.75±0.02	8.8– 9.1	2	10
		6.7	33						
4477	Σ 1196 AB	5.3	18 1	1889.19	42.4±1.6	1.02±0.02	5.2– 5.4	2	10
		9.3	17 49	1907.21	342.1±0.3	1.02±0.04	5.0– 5.1	4	10½
				1918.29	290.9±0.5	0.81±0.02	5.0– 5.4	4	10½
				1919.18	287.0±1.4	0.79±0.03	5.0– 5.4	4	10½
				1920.18	281.8±0.6	0.80±0.03	5.0– 5.4	4	10½
				1921.25	274.0±0.5	0.73±0.02	5.0– 5.6	4	10½
				1922.24	264.4±0.5	0.78±0.04	5.0– 5.5	4	10½
				1923.27	257.6±0.2	0.75±0.04	5.0– 6.0	2	10½
				1924.22	249.7±1.2	0.74±0.02	5.0– 5.5	2	10½
				1925.23	240.6±0.7	0.75±0.02	5.0– 5.6	5	10½
				1926.21	229.3±0.7	0.65±0.03	5.0– 5.4	5	10½
		8ʰ 9.3ᵐ	17°49′	1927.31	218.9	0.63	. .–. . .	1	10½

BGC	Double Star	R. A. 1880 / 1950	Dec. 1880 / 1950	Epoch	Position Angle	Distance	Magnitudes	Nights Observed	Apt.
4477	Σ 1196 AC	8ʰ 9.3ᵐ	17°49′	1889.19	120.4±0.5	5.66±0.03	5.3– 5.6	2	10
				1907.20	112.4±0.4	5.34±0.10	. .–. . .	2	10½
4477	Σ 1196 AB & C			1907.21	114.4±0.7	5.56±0.15	. .–. . .	4	10½
				1918.29	111.5±0.3	5.17±0.10	. .– 5.9	4	10½
				1919.18	110.6±0.5	5.52±0.08	. .– 5.6	4	10½
				1920.18	108.4±0.3	5.56±0.04	. .– 5.6	4	10½
				1921.25	109.4±0.5	5.56±0.09	. .– 6.0	4	10½
				1922.23	107.9±0.1	5.64±0.06	. .– 5.8	3	10½
				1923.27	108.0±0.1	5.36±0.08	. .–. . .	2	10½
				1924.22	107.6±1.0	5.44±0.02	. .– 5.8	2	10½
				1925.23	106.2±0.6	5.38±0.03	. .– 5.8	5	10½
				1926.21	107.4±1.0	5.32±0.10	. .– 5.9	6	10½
				1927.31	105.2	5.25	. .–. . .	1	10½
4477	Σ 1196 BC			1889.19	129.3±0.7	5.39±0.03	5.4– 5.6	2	10
				1907.19	116.6	5.88	. .–. . .	1	10½
4494	β 204	7.0 / 10.8	10 45 / 33	1889.11	304.2	1.00	7.8– 9.4	1	10
				1916.27	297.6±0.6	1.00±0.06	7.2– 9.9	2	10½
4507	β 904	7.9 / 11.4	− 5 23 / 35	1892.15	79.5±0.3	3.48±0.20	8.0–11.4	2	10
4537	β 905	11.0 / 14.2	−15 57 / −16 10	1888.57	11.7±0.7	4.07	8.1–10.4	2–1	10
4538	β 102	11.0 / 14.4	− 8 39 / 52	1916.24	117.4±1.4	3.31±0.14	7.1–10.8	3	10½
4539	β 454	11.1 / 13.9	−30 33 / 46	1892.26	16.6±0.8	2.49±0.15	7.8– 9.8	2	10
4668	β 205	27.9 / 30.9	−24 12 / 26	1889.16	258.1	0.71	7.0– 7.2	1	10
				1917.25	194.8±4.8	0.66±0.04	7.0– 7.0	5	10½
				1918.28	197.8	0.59	. .–. . .	1	10½
				1921.28	189.7	0.51	7.0– 7.2	1	10½
. . .	Barnard . . .	28.6 / 32.6	20 2 / 19 48	1917.51	163.5±1.2	4.80±0.25	10.4–11.1	4–3	10½
4684	β 206	30.3 / 33.3	−24 42 / 56	1886.21	285.2	. . . . . .	. .–. . .	1–0	26
				1889.10	280.3±0.4	1.72±0.01	7.9– 8.4	2	10
				1916.23	277.9±1.0	1.66±0.05	7.9– 8.4	3	10½
4708	β 207	33.3 / 36.5	−19 19 / 33	1888.65	102.4±1.0	4.05±0.11	7.0–10.2	2	10
4714	β 208	33.9 / 8ʰ 37.0ᵐ	−22 16 / −22°31′	1917.24	195.8±1.0	1.08±0.02	6.2– 8.9	5	10½
				1918.28	199.2±2.7	1.26±0.07	6.0– 9.0	3	10½

BGC	Double Star	R. A. 1880 1950	Dec. 1880 1950	Epoch	Position Angle	Distance	Magnitudes	Nights Observed	Apt.
4714	β 208	8ʰ 37.0ᵐ	−22°31′	1919.18	199.9	1.17	6.0– 8.5	1	10½
				1921.28	199.8	1.35	6.0– 7.4	1	10½
4743	Σ 1263	37.3	42 8	1891.96	20.5	49.02	. .–. . .	1	10
		42.0	41 53						
4771	Σ 1273 AB&C	40.4	6 52	1889.15	230.8±0.1	3.20±0.02	3.0– 7.0	2	10
		44.1	37	1907.28	235.8±0.8	3.30±0.07	3.0– 7.1	6	10½
				1920.21	245.1±0.5	3.05±0.19	3.5– 6.5	5	10½
				1921.28	245.3±0.5	3.29±0.04	3.0– 7.0	3	10½
				1922.32	245.3±0.3	3.14±0.05	3.0– 6.8	3	10½
				1923.37	247.5±0.8	3.10±0.02	3.0– 7.0	2	10½
				1925.40	249.1	3.53	. .–. . .	1	10½
				1926.29	248.6±0.5	3.23±0.21	3.0– 7.0	5	10½
4771	Σ 1273 AB			1920.19	340.4	0.4	. .–. . .	1	10½
. . .	Roe 72	43.7	31 5	1917.28	190.0±0.6	6.48±0.05	9.6–10.9	4	10½
		48.0	30 50						
4832	Schj 11	46.1	−10 41	1886.14	349.9	2.66	. .–. . .	1	26
		49.9	56	1922.29	354.8±0.2	2.15±0.03	8.7– 8.7	2	10½
4839	Σ 1291	46.9	31 2	1920.28	322.7±1.4	1.43±0.06	6.0– 6.2	3	10½
		51.2	30 46	1922.50	322.0±0.8	1.52±0.09	6.0– 6.2	4	10½
				1924.55	322.8±1.6	1.51±0.06	6.0– 6.1	3	10½
				1926.22	322.3±1.0	1.47±0.02	6.0– 6.2	3	10½
4849	β 24	48.4	− 8 18	1888.64	176.2±0.1	1.20±0.00	7.7– 8.8	2	10
		51.8	34						
4867	β 210	51.3	−16 58	1886.19	182.2	2.72	. .–. . .	1	26
		54.5	−17 14	1888.15	181.8±0.0	2.80±0.05	6.2– 6.6	3	10
				1916.22	181.7±0.1	2.99±0.06	7.0– 7.2	2	10½
				1921.27	181.8±0.5	3.04±0.01	7.0– 7.2	2	10½
4901	β 211	55.7	3 9	1888.58	262.8±0.2	1.02±0.00	7.1– 9.1	2	10
		59.3	2 53	1916.25	265.6±1.4	1.07±0.04	7.5– 9.7	3	10½
4923	Σ 1306	8 59.8	67 37	1897.29	208.7±2.0	1.24±0.02	5.0– 9.0	2	10½
		9 6.1	21						
4941	Σ 1316 AB	1.9	− 6 39	1886.16	140.0±0.7	6.82±0.12	. .–. . .	3–2	26
		5.4	56						
4941	Σ 1316 AC			1886.16	166.5±0.6	8.13±0.03	. .–. . .	3–2	26
4941	Σ 1316 BC	9ʰ 5.4ᵐ	− 6°56′	1886.16	45.7±1.1	3.56±0.15	. .–. . .	3–2	26

BGC	Double Star	R. A. 1880 1950	Dec. 1880 1950	Epoch	Position Angle	Distance	Magnitudes	Nights Observed	Apt.
4958	β 410	9ʰ 4.5ᵐ 7.6	−25°19′ 36	1892.22	161.3±0.0	1.74±0.12	7.5– 9.2	2	10
4966	β 104	5.3 8.9	0 47 30	1916.25	101.8±0.9	3.21±0.13	7.5–11.6	4–3	10½
4990	β 455	8.6 12.3	4 43 26	1888.83 1907.30	71.2±1.7 66.2±0.6	1.90±0.03 1.76±0.0	8.2–10.2 8.0–11.0	3 2–1	10 10½
5001	β 212	10.2 13.6	− 7 51 − 8 8	1887.02 1888.21 1916.46 1920.29 1921.26	224.7 226.4±0.1 217.2±0.6 218.3 216.0±0.2	1.49 1.27±0.00 1.29±0.10 . . . . . . 1.36	7.3– 8.0 7.9– 8.1 8.0– 8.4 . .–. . . 8.0– 8.6	1 2 4 1–0 2–1	26 10 10½ 10½ 10½
5005	Σ 3121	10.8 15.0	29 5 28 48	1907.30 1918.28	35.6±2.05 31.4	0.62±0.11 0.57	7.0– 7.0 . .–. . .	2 1	10½ 10½
5030	Σ 1338	13.5 17.9	38 42 25	1918.34 1920.28 1922.27 1924.22 1925.21 1926.33	178.4±1.3 181.8±0.9 183.8±0.4 185.4±0.2 186.6±0.8 187.2±1.4	1.37±0.09 1.42±0.09 1.58±0.10 1.39±0.09 1.34±0.06 1.46±0.02	7.1– 7.4 7.0– 7.3 7.0– 7.3 7.0– 7.2 7.0– 7.6 7.0– 7.4	4 3 3 2 2 4	10½ 10½ 10½ 10½ 10½ 10½
5033	Σ 1343	13.7 17.4	5 31 14	1888.27 1921.30	272.0 272.9	9.90 9.71	8.8– 9.0 8.5– 8.6	1 1	10 10½
5057	β 337	16.9 20.2	−17 23 41	1892.22 1907.27	326.6±0.2 330.1±0.5	7.78±0.06 8.36±0.16	7.0–10.2 7.8–10.8	2 3–2	10 10½
5062	β 105	17.7 21.8	26 42 24	1918.27	206.6±0.8	2.94±0.06	5.2–10.7	2	10½
5071	Σ 1348	18.2 21.9	6 52 34	1888.49 1907.24 1920.22 1922.32 1925.27 1926.31	325.2±0.7 322.8±0.2 321.3±0.4 319.4±0.3 321.3±0.2 319.9±1.2	1.68±0.08 1.79±0.03 2.02±0.06 2.04±0.01 1.78±0.13 1.75±0.01	7.2– 7.3 7.0– 7.0 7.1– 7.0 7.0– 7.0 7.0– 7.1 7.0– 7.0	3 3 2 2 3 2	10 10½ 10½ 10½ 10½ 10½
5094	Σ 1355	21.0 24.7	6 46 28	1889.08 1907.20 1920.22 1922.30	332.3 335.1±0.3 337.0±1.0 337.2	2.80 2.67±0.01 2.45±0.04	7.0– 7.0 7.0– 7.0 7.2– 7.3 . .–. . .	1 2 2 1–0	10 10½ 10½ 10½
5103	Σ 1356	22.0 25.7 9ʰ 25.7ᵐ	9 35 17 9°17′	1888.12 1889.05 1890.88 1892.35	94.4±0.8 97.6±0.8 101.5 105.0±1.0	0.72±0.04 0.68±0.01 0.64 0.72±0.03	6.1– 6.8 5.9– 7.1 6.5– 7.0 5.8– 6.6	3 4 1 2	10 10 10 10

BGC	Double Star	R. A. 1880 1950	Dec. 1880 1950	Epoch	Position Angle	Distance	Magnitudes	Nights Observed	Apt.
5103	Σ 1356	9ʰ 25.7ᵐ	9°17′	1907.29	121.9±0.4	0.90±0.05	6.0– 6.8	3	10½
				1918.28	127.3±2.5	0.99±0.03	6.0– 7.0	4	10½
				1919.35	127.8	1.06	6.0– 7.0	1	10½
				1920.24	129.0±1.7	0.97±0.04	6.0– 6.8	4	10½
				1921.29	129.1±0.5	0.97±0.06	6.0– 6.8	3	10½
				1922.30	131.0±0.8	1.08±0.07	6.0– 6.9	3	10½
				1923.34	132.2±1.3	1.18±0.08	6.0– 6.7	4	10½
				1925.23	133.2±1.2	1.14±0.05	5.8– 6.3	5	10½
				1926.21	134.7±1.3	1.08±0.05	6.0– 6.5	5	10½
5106	β 213	22.4	— 7 34	1916.27	176.0±1.2	1.24±0.04	8.1–10.8	2	10½
		25.9	52	1917.21	174.3	1.47	8.0–10.5	1	10½
5126	β 339	25.3	—15 13	1888.23	220.8	1.30	9.0–10.2	1	10
		28.6	31						
5181	β 214	35.9	—17 56	1886.29	262.0	3.22	8.0–10.5	1	26
		39.2	—18 15						
5187	Σ 1377	37.2	3 11	1888.15	139.0	. . . . . .	7.5–10.5	1–0	10
		40.8	2 52	1921.30	140.8	. . . . . .	. .–. . .	1–0	10½
5223	OΣ 208	43.9	54 38	1921.38	324.8	0.54	5.0– 5.5	1	10½
		48.7	19	1922.31	325.0±1.8	0.49±0.01	5.2– 5.0	3	10½
				1923.35	325.8±0.7	0.47±0.03	5.0– 5.7	3	10½
				1925.36	328.9±1.6	0.51±0.02	5.0– 5.5	4	10½
				1926.32	330.8±0.3	0.53±0.02	5.0– 5.6	6	10½
5235	A. Clark 5	46.6	— 7 32	1888.12	133.1	0.4	. .–. . .	1	10
		50.1	52	1918.31	62.4±1.0	0.59±0.04	5.0– 5.5	5	10½
				1920.26	61.2	0.65	5.0– 5.0	1	10½
				1921.32	59.9±0.4	0.62±0.02	5.1– 5.6	4	10½
				1922.32	56.6±1.1	0.62±0.01	5.3– 5.8	3	10½
				1923.34	56.5	0.62	5.0– 5.4	1	10½
				1925.31	55.8±0.3	0.64±0.01	. .–. . .	4	10½
				1926.28	54.7±0.8	0.59±0.02	5.0– 5.6	3	10½
5244	β 215	48.7	—27 26	1917.25	345.7±1.4	1.58±0.04	7.5–10.4	2–3	10½
		51.8	46						
5263	β 216	51.3	—25 59	1917.25	157.6±0.7	3.55±0.11	6.2–11.3	4	10½
		9 54.5	—26 19	1918.28	155.2	3.47	6.0–11.3	1	10½
5319	Hd 125	10 0.8	— 1 8	1888.21	172.1±0.3	2.02±0.05	8.8– 9.0	3	10
		4.3	28	1922.28	178.0±0.0	2.24±0.04	8.7– 8.7	2	10½
				1925.30	179.6±0.4	2.20±0.16	8.8– 8.9	4	10½
				1926.31	178.2±0.8	2.48±0.24	8.8– 8.8	2	10½
5325	β 217	1.3	—24 18	1921.32	289.2±1.8	1.78±0.08	8.1– 8.0	4	10½
		10ʰ 4.5ᵐ	—24°38′						

BGC	Double Star	R. A. 1880 / 1950	Dec. 1880 / 1950	Epoch	Position Angle	Distance	Magnitudes	Nights Observed	Apt.
5329	β 218	10^h 1.7^m / 5.0	—19° 7′ / 27	1888.11 / 1921.31	120.0±0.6 / 125.5±1.3	0.87±0.09 / 0.86±0.11	7.5– 7.9 / 8.0– 8.6	2 / 3	10 / 10½
. . .	Fox . . .	2.5 / 6.4	25 36 / 15	1917.63	70.6±1.3	7.87±0.67	10.8–11.6	3	10½
5339	β 790	4.1 / 7.5	—12 17 / 37	1888.96	68.5±1.5	2.11±0.06	9.3–10.5	4–3	10
5359	Σ 1417	8.6 / 12.4	19 43 / 22	1888.74	259.6±0.8	2.42±0.10	8.4– 8.4	2	10
				1907.22	260.0±0.2	2.46±0.04	9.0– 9.0	2	10½
				1920.28	258.0±0.2	2.25±0.01	8.5– 8.6	2	10½
5365	OΣ 215	9.7 / 13.5	18 20 / 17 59	1892.39	212.3±0.1	0.66±0.08	7.0– 7.4	2	10
				1907.24	204.1±0.5	0.83±0.05	7.0– 7.4	3	10½
				1917.39	200.7±0.9	0.91±0.06	7.0– 7.4	3	10½
				1918.30	199.8±0.1	0.98±0.04	7.0– 7.6	4	10½
				1919.39	201.8±1.1	0.97±0.01	7.0– 7.3	3	10½
				1920.25	200.9±0.4	0.96±0.07	7.0– 7.4	3	10½
				1921.32	199.7±0.6	1.01±0.05	7.0– 7.3	4	10½
				1922.26	199.8±0.2	0.99±0.05	7.0– 7.2	2	10½
				1923.32	201.3±0.1	1.06±0.10	7.0– 7.3	2	10½
				1925.22	198.6±0.2	1.03±0.02	7.0– 7.3	2	10½
				1926.25	197.7±0.8	1.10±0.03	7.0– 7.2	3	10½
. . .	Espin 915	12.7 / 17.1	48 20 / 47 59	1917.81	165.2±0.6	5.84	9.5–12.0	2–1	10½
5388	Σ 1424	13.3 / 17.1	20 27 / 6	1892.36	113.9±0.7	3.45±0.04	. .–. . .	2	10
				1907.31	116.4±0.3	3.67±0.07	2.0– 3.5	3	10½
				1914.36	118.2±0.4	3.95±0.09	2.5– 4.8	3	10½
				1918.35	116.8±0.5	3.80±0.06	3.0– 4.3	5	10½
				1919.42	118.4±0.8	3.71±0.10	2.0– 4.0	4	10½
				1920.34	118.7±0.3	3.74±0.02	2.0– 3.1	3	10½
				1921.33	118.2±0.1	3.80±0.06	2.0– 3.1	4	10½
				1922.35	118.6±0.2	3.86±0.01	2.0– 3.4	4–3	10½
				1923.35	118.3±0.1	3.83±0.03	2.0– 3.4	3	10½
				1925.35	118.4±0.6	3.71±0.04	2.5– 3.5	11	10½
				1926.30	117.8±0.7	3.71±0.04	2.5– 3.8	10	10½
5407	β 25	15.8 / 19.3	— 9 10 / 31	1886.30	178.9	1.96	8.0– 9.0	1	26
				1888.22	178.5±0.7	1.86±0.08	8.0– 8.8	3–2	10
				1907.24	170.8±0.5	1.74±0.05	8.0– 9.3	4	10½
				1916.23	167.6±0.2	1.58±0.01	8.0– 8.9	3	10½
				1921.34	168.0±0.0	1.68±1.20	8.3– 8.8	2	10½
				1922.32	167.3±0.7	1.75±0.08	8.1– 8.8	3	10½
				1925.36	166.3±0.1	1.68±0.06	8.0– 8.8	2	10½
				1926.32	166.1±0.1	1.64±0.10	8.0– 9.0	2	10½
5408	β 219	15.9 / 10^h 19.2^m	—21 55 / —22°16′	1892.20	190.2±2.3	2.30±0.16	7.0– 9.2	2	10
				1916.25	185.0±0.5	2.19±0.18	7.2– 9.1	3–2	10½

BGC	Double Star	R. A. 1880 1950	Dec. 1880 1950	Epoch	Position Angle	Distance	Magnitudes	Nights Observed	Apt.
5414	h 4311	10ʰ 17.4ᵐ 20.8	—12°46′ —13 7	1886.14	126.2±0.4	3.98±0.07	. .—. . .	1	26
5437	OΣ 218	21.3 24.9	4 10 3 49	1888.92 1922.33 1926.30	70.4±1.4 76.6±0.8 80.9±0.6	1.04±0.03 1.01±0.40 1.03±0.10	7.1– 9.1 6.8– 9.2 7.0– 9.3	4–2 3 3	10 10½ 10½
5453	h 4321	24.1 27.3	—30 0 21	1888.65	225.6±0.0	11.03±0.02	5.5– 8.5	2	10
. . .	Jon 736	24.0 27.8	15 21 0	1917.31 1918.29	201.6±1.2 202.6±1.3	1.54±0.26 1.97±0.41	10.0–10.4 9.7–10.2	2 3	10½ 10½
5459	Σ 1441	25.0 28.5	— 7 1 22	1886.29 1921.34	165.5 167.0±0.6	2.87 3.00±0.10	6.0–10.0 6.0– 9.8	1 2	26 10½
5484	Σ 1450	28.8 32.5	9 16 8 54	1889.16 1921.32	157.8 156.9±0.5	2.48 2.50±0.09	6.0– 9.0 6.0– 9.0	1 2	10 10½
5491	β 411	30.4 33.7	—26 3 25	1892.30	289.8±0.5	1.10±0.07	7.0– 8.5	2	10
5508	Σ 1457	32.5 36.1	6 21 5 59	1888.23 1907.30 1920.34 1921.29 1922.28 1923.34 1925.32 1926.27	318.0±0.7 319.4±0.2 324.5 322.2±0.5 323.0±0.4 322.8 323.1±1.0 322.2±1.4	1.20±0.09 1.26±0.03 1.42 1.47±0.09 1.57±0.08 1.46 1.63±0.08 1.56±0.05	8.0– 8.2 7.9– 8.3 8.0– 8.5 7.5– 7.8 8.1– 8.4 7.5– 7.8 7.8– 8.2 7.4– 7.9	2 3 1 3–2 3 1 3 4	10 10½ 10½ 10½ 10½ 10½ 10½ 10½
. . .	Espin 603	35.5 39.7	48 49 27	1918.31	97.9	12.83	9.3–11.0	1	10½
. . .	Espin 1246	35.5 39.6	43 41 19	1918.31	55.3	3.40	9.3–10.0	1	10½
5557	Σ 1472	40.6 44.3	13 40 18	1917.36	38.5±0.3	38.41±0.11	8.5– 9.0	3	10½
. . .	AG . . .	41.8 45.5	13 50 28	1917.36	231.0±0.4	36.79±0.38	9.3–10.1	3	10½
5572	Σ 1476	43.2 46.7	— 3 23 45	1888.19 1907.26 1921.26 1926.28	359.7±0.1 1.7±0.1 5.1±0.0 5.3±0.4	2.20±0.01 2.26±0.02 2.01±0.12 2.29±0.13	7.0– 7.8 7.1– 7.1 7.0– 7.4 7.0– 7.6	2 2 2 3	10 10½ 10½ 10½
5573	β 915	43.2 10ʰ 47.0ᵐ	24 55 24°33′	1892.30	232.5±4.4	0.7 ±0.1	9.0– 9.0	4	10

BGC	Double Star	R. A. 1880 1950	Dec. 1880 1950	Epoch	Position Angle	Distance	Magnitudes	Nights Observed	Apt.
5579	β 111	10^{h}45.2^m	— 8°28′	1886.30	4.9	3.55	. .—. . .	1	26
		48.7	50	1888.24	5.4±1.4	3.64±0.20	9.4– 9.5	3	10
				1907.26	3.7±1.0	4.04±0.03	9.3– 9.2	3	10½
				1922.31	3.7	3.61	9.0– 9.3	1	10½
. . .	Espin 433	50.6	30 22	1917.36	221.2±0.5	5.87±0.26	9.6–11.4	3–2	10½
		54.4	0						
. . .	Arg 73	52.4	9 46	1917.33	70.6±1.1	16.63±0.17	9.4– 9.6	4	10½
		56.1	24						
. . .	AG . . .	53.2	26 5	1917.32	79.5±1.3	4.90±0.21	8.4– 8.7	4	10½
		57.0	25 43						
. . .	A 1772	53.6	—13 52	1918.33	212.3±0.1	3.96±0.12	9.4–10.6	2	10½
		7.0	—14 14						
5650	H I 77	56.2	—15 8	1888.19	17.1±0.3	2.94±0.08	8.1– 8.2	2	10
		59.7	30	1922.32	18.8±0.4	2.94±0.04	8.0– 8.2	2	10½
5656	Howe 25	10 57.5	—26 52	1889.24	332.7	2.14	8.3– 9.5	1	10
		11 0.9	—27 15	1921.31	339.7±1.7	2.13±0.12	7.9– 9.5	3	10½
5659	Σ 1504	10 57.8	4 17	1888.18	283.0±0.3	1.18±0.05	7.2– 7.1	3	10
		11 1.4	3 54	1907.32	288.2±1.2	1.15±0.06	7.5– 7.6	3	10½
				1917.38	289.2±0.6	1.15±0.03	7.7– 7.5	3	10½
5702	β 220	6.6	—17 51	1889.09	138.4±7.0	0.70±0.02	6.1– 6.8	4–2	10
		10.1	—18 14	1917.06	157.5±10.3	0.50±0.05	7.2– 7.2	3–4	10½
				1918.32	163.3±9.1	0.51±0.02	. .—. . .	2	10½
				1921.80	155.8±6.3	0.41±0.01	. .—. . .	2	10½
				1922.33	162.1	. . . . . .	. .—. . .	1–0	10½
5710	β 916	8.1	—14 47	1888.45	358.4±2.4	0.65	7.0– 8.2	3–1	10
		11.6	—15 10	1922.33	5.6	. . . . . .	. .—. . .	1–0	10½
				1926.29	8.9±5.0	0.52±0.02	7.5– 8.0	2	10½
5734	Σ 1523	11.8	32 13	1892.45	196.6±0.6	1.60±0.11	4.8– 5.5	2	10
		15.6	31 50	1907.33	131.0±1.7	2.69±0.05	4.0– 4.6	4	10½
				1919.43	106.8±0.8	2.77±0.01	4.0– 4.4	2	10½
				1920.40	103.6±0.2	2.86±0.06	4.0– 4.5	4	10½
				1921.38	101.4±0.4	2.70±0.05	4.0– 4.5	5	10½
				1922.37	98.7±0.5	2.74±0.02	4.0– 4.4	3	10½
				1923.41	96.0±0.5	2.54±0.04	4.0– 4.6	5	10½
				1924.53	93.7±0.6	2.46±0.05	4.0– 4.4	3	10½
				1925.41	89.6±0.5	2.30±0.16	4.0– 4.3	6	10½
				1926.38	86.2±0.9	2.11±0.10	4.0– 4.6	9	10½
		11^{h}15.6^m	31°50′	1927.39	79.2±0.5	1.93±0.05	4.0– 4.4	3	10½

BGC	Double Star	R. A. 1880 1950	Dec. 1880 1950	Epoch	Position Angle	Distance	Magnitudes	Nights Observed	Apt.
5739	Σ 1527	11ʰ 12.7ᵐ	14°56′	1889.26	14.9±0.3	3.42±0.09	7.0– 7.6	2	10
		16.4	33	1907.32	16.3±0.7	3.15±0.08	7.3– 8.0	3	10½
				1920.34	18.9±0.4	3.02±0.02	7.0– 7.6	3	10½
				1926.29	19.8±1.3	2.80±0.02	7.0– 7.9	4	10½
				1927.36	19.9±0.5	2.83±0.06	7.0– 7.8	3	10½
5748	Σ 1530	13.7	− 6 15	1886.29	313.1	7.72	8.5– 8.5	1	26
		17.2	38	1920.34	313.2±0.5	7.67±0.11	8.0– 8.2	3	10½
5765	Σ 1536	17.6	11 12	1889.17	63.5±1.4	2.56±0.01	4.3– 7.4	2	10
		21.2	10 49	1907.39	47.5±0.8	2.21±0.04	4.0– 7.0	4	10½
				1914.37	42.0±0.6	2.10±0.08	4.0– 7.0	4	10½
				1918.30	37.1±0.6	2.06±0.08	4.0– 7.0	4	10½
				1919.39	37.9±0.7	1.77±0.22	4.0– 7.3	5	10½
				1920.35	37.4±0.9	1.72±0.09	4.0– 7.3	4	10½
				1921.35	35.6±0.8	1.80±0.02	4.0– 7.0	4	10½
				1922.38	33.2±0.4	1.57±0.13	4.0– 7.2	4	10½
				1923.34	32.1±1.3	1.74±0.04	4.0– 7.0	4	10½
				1925.32	27.6±1.5	1.45±0.11	. .−. . .	8–7	10½
				1926.33	28.9±1.5	1.40±0.07	. .−. . .	9	10½
				1927.40	25.5±0.9	1.26±0.03	. .−. . .	2	10½
5766	β 26	17.7	− 9 46	1886.30	67.4	2.94	. .−. . .	1	26
		21.2	−10 9	1888.67	69.8±0.5	2.84±0.03	7.9–10.4	2	10
				1907.33	66.2	2.90	8.0–10.5	1	10½
				1914.35	68.5	2.76	8.0–10.0	1	10½
				1921.34	67.2±0.3	3.14±0.15	8.0–11.0	2	10½
5796	β 601 BC	23.2	−16 41	1886.29	220.4	. . . . . .	. .−. . .	1	26
		26.7	−17 4	1888.65	224.8±0.7	0.75±0.5	7.7– 8.7	2	10
				1922.33	212.9	0.77	8.0– 9.0	1	10½
5805	OΣ 234	24.3	41 57	1892.45	136.5	0.4	. .−. . .	1	10
		28.1	34	1907.33	148.9±1.7	0.49	. .−. . .	5	10½
				1921.41	169.4	0.45	7.0– 7.5	1	10½
				1922.38	163.5	0.50	7.0– 7.3	1	10½
				1923.38	166.2±3.0	0.48±0.02	. .−. . .	2	10½
				1925.36	172.1±3.0	0.49±0.01	7.0– 7.2	3	10½
				1926.36	177.8	0.47	7.0– 7.2	1	10½
5811	OΣ 235	25.5	61 45	1892.45	85.4±1.2	0.80±0.02	5.2– 7.0	2	10
		29.5	22	1921.41	310.4	0.56	6.0– 7.5	1	10½
5855	Σ 1560	32.2	− 1 46	1885.33	279.3	5.56	. .−. . .	1	26
		35.8	− 2 9	1922.27	277.0±0.6	5.03±0.17	6.0–10.4	3	10½
				1925.69	277.0±0.3	5.11±0.13	6.0–10.3	3	10½
5888	β 917	37.4	11 22	1892.29	178.3	3.38	8.0–11.0	1	10
		11ʰ 41.0ᵐ	10 59′						

BGC	Double Star	R. A. 1880 1950	Dec. 1880 1950	Epoch	Position Angle	Distance	Magnitudes	Nights Observed	Apt.
5926	β 603	11ʰ42.5ᵐ 46.1	14°57′ 34	1889.27	327.5	. . . . . .	7.0–10.0	1–0	10
6002	β 457	55.2 58.8	—20 52 —21 15	1886.30	82.3	1.25	8.0–10.0	1	26
6013	Σ 1593	11 57.4 12 1.0	— 1 47 — 2 10	1885.35 1922.26	17.9 17.2±0.6	1.92 1.14±0.02	. .–. . . 8.2– 8.3	1 2	26 10½
. . .	h 4495	0.0 3.5	—32 17 41	1889.22	314.5±0.4	6.59±0.01	6.8– 8.9	2	10
6041	β 412	2.2 5.8	—17 55 —18 18	1886.30 1907.30 1922.32	161.0 160.6±0.6 159.6±0.2	1.84 1.90±0.03 2.02±0.04	. .–. . . 8.2– 9.2 8.0– 9.0	1 3 2	26 10½ 10½
. . .	Espin 725	4.0 7.6	47 15 46 52	1917.39	58.3±0.2	6.85±0.39	9.5–10.0	4	10½
6094	β 920	9.6 13.2	—22 41 —23 4	1892.35 1907.35	250.0±2.5 266.9	0.81±0.03 0.77	6.5– 7.8 . .–. . .	4–3 1	10 10½
6095	Σ 1620	9.7 13.3	9 42 19	1892.33	81.5	1.93	8.5–10.0	1	10
6112	β 921	11.7 15.3	—23 21 44	1892.34	218.5±1.3	2.95±0.07	7.0–12.0	4–3	10
6127	β 605	14.0 17.6	—21 30 53	1889.04	137.5	. . . . . .	6.0– 8.5	1–0	10
6129	β 27	14.0 17.6	14 31 8	1916.27	103.7±0.6	3.31±0.06	7.5–11.3	3–4	10½
6158	Σ 1639	18.4 21.9	26 15 25 52	1907.38 1917.40 1918.33 1921.35 1922.29 1923.36 1924.56 1925.36 1926.34 1927.41	347.8±1.6 344.7 342.9 338.7±0.8 338.6±1.1 339.4±1.4 338.3 338.5±1.8 337.0±2.4 337.1	0.42±0.04 0.57 0.64 0.81±0.03 0.90±0.07 0.86±0.10 0.92 0.81±0.08 0.88±0.05 0.84	. .–. . . 6.0– 7.5 6.0– 6.7 7.0– 8.0 6.9– 7.9 7.0– 8.1 7.0– 8.5 7.0– 7.9 7.0– 8.3 . .–. . .	2 1 1 2 3 3 1 7–6 3 1	10½ 10½ 10½ 10½ 10½ 10½ 10½ 10½ 10½ 10½
6168	Hn 13	20.3 23.9	— 1 13 36	1888.31 1907.38	155.2±2.2 148.1±0.5	1.28 1.25±0.00	8.7– 9.2 9.1– 9.6	2–1 2–1	10 10½
. . .	Doo 164	22.4 12ʰ26.0ᵐ	5 11 4°48′	1920.37	219.2	2.48	9.8–10.2	1	10½

BGC	Double Star	R. A. 1880 1950	Dec. 1880 1950	Epoch	Position Angle	Distance	Magnitudes	Nights Observed	Apt.
6185	β 28	12ʰ 23.9ᵐ	−12°44′	1885.34	2.9	2.20	. .−. . .	1	26
		27.5	−13 7	1889.24	3.0±0.5	. . . . . .	. .−. . .	2–0	10
				1916.29	27.6±1.4	1.29±0.04	6.0–10.5	2	10½
				1917.40	23.6	1.08	6.0–11.0	1	10½
				1918.33	30.2	1.54	6.0–10.0	1	10½
				1922.32	33.8±0.4	1.48	6.0–10.5	2–1	10½
6185	β 28 AC			1918.34	293.8±0.4	90.99±0.41	7.0–11.1	4–3	10½
6185	β 28 AD			1918.34	183.3±1.1	79.17±0.18	7.0–12.1	3–2	10½
6187	Σ 1647	24.5	10 23	1888.30	222.8±0.2	1.43±0.06	7.9– 8.2	2	10
		28.1	0	1907.35	222.4±0.6	1.20±0.07	8.0– 8.2	2	10½
				1920.38	228.0±0.7	1.29±0.07	8.0– 8.6	3	10½
				1922.38	228.8±0.4	1.47±0.01	7.8– 8.1	2	10½
				1925.33	229.5±0.1	1.37±0.01	7.9– 8.1	2	10½
				1926.39	229.4±0.1	1.44±0.02	8.0– 8.7	2	10½
6208	Lv 5	27.9	−17 32	1886.30	33.5±0.5	1.70±0.24	7.6– 9.8	2	26
		31.5	55	1888.19	34.1±1.5	1.41	7.1– 9.7	2–1	10
				1921.31	14.5±2.1	1.12	7.0–10.0	2–1	10½
				1923.03	17.6±2.8	1.39±0.01	7.2–10.5	3–2	10½
				1925.35	18.8±0.2	1.16±0.01	7.0–10.0	2	10½
6214	h 1218	29.5	−16 10	1916.29	261.9±1.2	11.53±0.10	6.8–11.0	3	10½
		33.1	33						
. . .	Espin 924	29.0	48 48	1917.40	220.9	3.96	9.5–10.2	1	10½
		32.3	25						
. . .	AG . . .	32.8	13 31	1917.39	325.5±0.4	8.53±0.18	9.2–11.0	4	10½
		36.3	8						
6236	Σ 1668	34.8	9 29	1889.27	195.4	1.46	8.0– 8.3	1	10
		38.3	6	1921.32	193.1±0.3	1.51±0.11	7.8– 8.2	3	10½
6239	Σ 1669	35.0	−12 21	1888.35	303.9±0.4	5.68±0.03	6.3– 6.4	3	10
		38.6	44	1907.41	305.8±0.5	5.70±0.03	5.9– 6.0	2	10½
				1922.39	307.1±0.5	5.44±0.04	6.0– 6.1	3	10½
				1925.39	307.1±0.3	5.32±0.07	6.0– 6.1	3	10½
				1926.38	306.2±0.8	5.36±0.03	6.1– 6.0	2	10½
6243	Σ 1670	35.6	− 0 47	1888.35	335.5±0.8	5.46±0.05	3.2– 3.2	4	10
		39.1	− 1 10	1889.14	333.2±0.1	5.60±0.02	3.0– 3.3	2	10
				1892.42	332.6	5.72	. .−. . .	1	10
				1911.43	326.3±0.2	6.00±0.06	3.0– 3.1	4	10½
				1913.45	325.3±0.3	6.00±0.04	3.5– 3.5	4	10½
				1914.40	325.7±0.3	6.06±0.06	3.6– 3.5	3	10½
		12ʰ 39.1ᵐ	− 1°10′	1917.40	324.0±0.1	5.94±0.02	3.0– 3.1	2	10½

BGC	Double Star	R. A. 1880 / 1950	Dec. 1880 / 1950	Epoch	Position Angle	Distance	Magnitudes	Nights Observed	Apt.
6243	Σ 1670	12ʰ 39.1ᵐ — 1°10′		1919.48	322.8±0.8	5.75±0.14	3.5– 3.5	4	10½
				1920.42	321.7±0.4	5.84±0.06	3.5– 3.7	3	10½
				1921.41	322.2±0.4	5.80±0.03	3.5– 3.6	4	10½
				1922.44	321.8±0.6	5.85±0.04	3.5– 3.6	5	10½
				1922.45	321.8±0.2	5.80±0.02	. .–. . .	2	10½
				1923.43	321.8±0.4	5.81±0.11	3.5– 3.6	5	10½
				1924.54	320.6±0.4	5.53±0.00	3.5– 3.5	3	10½
				1925.42	320.4±0.4	5.76±0.13	3.5– 3.7	6	10½
				1925.54	320.6±0.2	5.72±0.09	3.5– 3.6	4	10½
				1926.41	320.4±0.8	5.73±0.08	3.5– 3.7	8	10½
6296	Σ 1687	47.4	21 54	1917.44	92.6	1.25	5.0– 8.3	1	10½
		50.9	31	1919.50	99.3±0.4	1.16±0.02	5.0– 8.2	2	10½
				1920.36	96.6±1.3	1.13±0.04	5.0– 8.2	3	10½
				1921.42	96.9±1.0	1.14±0.06	5.0– 8.2	3	10½
				1922.34	96.5±1.2	1.23±0.07	5.0– 8.1	4	10½
				1924.47	98.6±0.5	1.11±0.24	5.0– 8.0	2	10½
6312	OΣ 256	50.3	— 0 18	1914.36	80.2±0.9	0.73±0.02	7.2– 7.0	2	10½
		53.9	41	1919.40	80.9±0.2	0.77±0.06	7.0– 7.2	3	10½
				1920.42	80.4±1.8	0.83±0.03	7.8– 7.0	2	10½
				1921.38	81.2±0.8	0.80±0.01	7.0– 7.1	2	10½
				1922.41	80.0±0.8	0.87±0.03	7.3– 7.0	2	10½
				1923.41	81.7±0.5	0.90±0.04	7.0– 7.1	2	10½
				1925.34	83.1±0.0	0.84±0.06	7.1– 7.0	2	10½
				1926.39	260.8±1.8	0.78	7.0– 7.2	2	10½
6313	Σ 1692	50.4	38 58	1926.58	227.8	19.80	. .–. . .	1	10½
		53.7	35						
6323	O. Stone 27	51.5	—12 29	1888.23	68.2±0.0	1.90±0.09	8.2– 8.2	2	10
		55.2	52	1921.85	67.6±0.2	2.15±0.06	8.2– 8.4	2	10½
6326	β 926	52.2	— 5 24	1892.32	268.0±0.2	2.24±0.10	8.0–11.8	2	10
		55.8	47						
6345	β 112	54.8	19 1	1892.32	291.1±2.1	1.90±0.04	9.1– 9.8	3	10
		58.3	18 38	1907.38	291.8±1.5	1.88±0.14	9.4– 9.8	2	10½
6345	β 112 A & BC			1907.38	349.7±0.0	150.6±0.2	. .–. . .	2	10½
6352	β 927	12 56.6	— 5 53	1892.32	290.5±1.1	4.42±0.07	8.2–10.0	3	10
		13 0.2	— 6 16	1922.44	289.2±0.6	4.72±0.14	8.6–10.5	2	10½
6362	β 928	12 57.2	— 5 47	1888.53	312.6±0.4	1.97±0.02	8.0– 9.3	3	10
		13 0.8	— 6 10	1922.44	312.5±0.5	2.12±0.10	8.1– 9.4	2	10½
6367	β 929	12 57.7	— 3 1	1888.32	221.6±1.5	0.62±0.01	6.0– 6.5	2	10
		13 1.3	24	1911.45	216.0±3.4	0.60±0.02	6.0– 6.6	2	10½
		13ʰ 1.3ᵐ — 3°24′		1914.02	211.6±0.8	0.64±0.06	6.1– 6.4	3	10½

BGC	Double Star	R. A. 1880 1950	Dec. 1880 1950	Epoch	Position Angle	Distance	Magnitudes	Nights Observed	Apt.
6367	β 929	13ʰ 1.3ᵐ	— 3°24′	1919.40	211.4±1.2	0.63±0.04	6.0– 6.4	2	10½
				1920.40	211.4	0.66	6.0– 6.8	1	10½
				1926.45	209.4±0.8	0.65±0.02	6.0– 6.9	3	10½
6406	Σ 1728	4.2	18 10	1889.08	10.5	0.56	5.6– 5.6	1	10
		7.6	17 48	1892.37	11.7±0.2	0.48±0.02	5.8– 6.2	2	10
				1907.37	190.4±1.2	0.57±0.04	6.0– 6.1	3	10½
				1914.36	189.2±1.2	0.57±0.00	5.0– 5.2	2	10½
				1917.46	192.4±0.3	0.49±0.00	5.5– 5.8	2	10½
				1919.40	191.2±2.1	0.42±0.02	6.2– 6.1	2	10½
6415	OΣ 261	6.4	32 43	1896.42	347.2±0.3	1.36±0.02	6.0– 6.5	2	10½
		9.7	21	1920.49	343.9±0.3	1.62±0.08	7.0– 7.2	2	10½
				1924.04	343.3±0.1	1.74±0.14	6.5– 6.8	2	10½
				1925.38	343.6±0.7	1.89±0.08	. .–. . .	3	10½
				1926.44	344.1±0.8	1.83±0.04	7.1– 7.2	3	10½
6421	β 221	6.9	—14 49	1886.30	45.6	1.70	. .–. . .	1	26
		10.6	—15 11	1916.32	48.3±0.3	1.56±0.3	8.0–10.6	2	10½
6422	Sh 151	7.0	—18 11	1926.44	34.1±0.4	5.15	7.0– 7.3	1	10½
		10.7	33						
6433	β 342	8.8	—18 17	1888.32	35.7±0.7	3.80±0.22	8.4– 8.7	2	10
		12.5	39	1926.44	34.4	4.24	8.5– 9.0	1	10½
6443	β 222	10.9	—20 54	1916.32	13.8±0.0	1.58±0.6	8.0– 9.4	2	10½
		14.7	—21 16						
6455	Σ 1734	14.6	3 34	1888.59	191.7±0.7	1.18±0.01	6.8– 7.6	3–2	10
		18.2	12	1907.41	187.9±0.7	1.02±0.00	7.0– 7.4	2	10½
				1917.46	188.6±0.6	0.98±0.04	7.0– 7.8	2	10½
				1926.44	187.4±0.4	1.18±0.06	7.0– 7.9	2	10½
6469	Σ 1738	16.8	—14 18	1886.33	281.8±0.3	4.35±0.14	. .–. . .	2	26
		20.5	40						
6473	β 610	17.5	—20 18	1892.38	18.5±0.6	3.82±0.15	7.1–11.8	3	10
		21.3	40						
6478	Σ 1742	18.2	2 2	1889.14	350.4±0.5	1.22±0.05	7.5– 7.7	2	10
		21.8	1 40	1923.37	353.0±0.8	1.22±0.06	7.5– 7.6	2	10½
6479	β 460	18.7	—15 0	1892.38	34.8±1.9	2.26±0.07	8.0–10.4	3	10
		22.4	22						
6482	Σ 1744	19.1	55 33	1926.58	149.7	14.51	2.2– 4.0	1	10½
		13ʰ21.9ᵐ	55°11′						

BGC	Double Star	R. A. 1880 1950	Dec. 1880 1950	Epoch	Position Angle	Distance	Magnitudes	Nights Observed	Apt.
6484	O. Stone 30	13ʰ 19.7ᵐ 23.5	−22°37′ 59	1886.30	355.1	1.85	9.0– 9.0	1	26
6500	β 113	23.2 26.7	12 6 11 44	1916.35 1922.42	216.6±2.9 222.7	1.24±0.04 1.30	8.4–11.2 8.5–10.5	5–3 1	10½ 10½
6528	β 114	28.0 31.7	− 8 0 22	1885.34 1888.49 1911.42 1919.40 1920.41	135.7 138.0±0.3 144.7±1.0 147.4±0.2 145.5±1.6	1.66 1.41±0.07 1.29±0.04 1.25±0.02 1.34±0.16	. .–. . . 8.1– 8.3 8.0– 8.4 8.2– 8.5 7.6– 8.2	1 3 3 2 2	26 10 10½ 10½ 10½
6530	Σ 1757 AB	28.2 31.8	+ 0 18 − 0 4	1888.59 1896.36 1911.42 1914.40 1919.40 1922.42 1923.49 1925.37	73.1±0.7 76.4 83.5±1.8 86.0±0.9 86.9±1.0 87.4±0.8 86.8 88.9±0.6	2.44±0.12 . . . . . . 2.71±0.10 2.82±0.16 2.71±0.08 2.68±0.10 2.62 2.58±0.15	7.9– 8.5 . .–. . . 8.0– 8.7 8.0– 8.4 8.0– 8.7 8.0– 8.7 7.8– 8.5 8.0– 8.6	3 1–0 5 4–3 3 2 1 4	10 10½ 10½ 10½ 10½ 10½ 10½ 10½
6530	Σ 1757 AC			1919.40 1922.38	151.4±0.4 149.5	43.48±0.80 43.64	. .–12.0 . .–12.0	2 1	10½ 10½
6541	β 933	29.1 32.3	33 45 23	1888.68 1923.44	29.2±0.6 28.5±0.9	2.07 2.16	8.2– 9.2 8.0– 9.3	2–1 2–1	10 10½
6566	Σ 1768	32.1 35.2	36 54 32	1911.48 1919.50 1922.42 1923.36 1924.58 1925.37 1926.51	128.9±0.2 123.3±1.6 120.0±1.4 119.4±1.2 119.6 118.4±1.5 119.2±1.4	1.30±0.05 1.65±0.08 1.56±0.10 1.54±0.06 1.53 1.68±0.04 1.60±0.12	5.0– 8.0 5.0– 7.7 5.0– 7.8 5.0– 7.5 . .–. . . 5.2– 7.4 5.0– 7.4	4 3 2 3 1 4 6	10½ 10½ 10½ 10½ 10½ 10½ 10½
6599	Σ 1777	37.0 40.5	4 9 3 48	1888.39 1922.34 1925.39 1926.46	234.9±0.9 230.2±0.4 230.8±0.6 231.2±0.2	3.46±0.07 3.25±0.05 3.27±0.07 3.31±0.23	5.8– 8.2 6.0– 8.5 6.0– 8.5 5.9– 8.6	3 3 3 2	10 10½ 10½ 10½
6610	β 223	39.0 42.6	− 2 43 − 3 4	1916.36	343.7±0.8	18.81±0.08	8.3–10.5	4	10½
6616	β 115 AB	39.4 42.9	9 40 19	1916.61 1918.32	233.8±2.5 238.9	1.52±0.11 1.35	8.0–11.3 . .–. . .	4–2 1	10½ 10½
6616	β 115 AC	13ʰ 42.9ᵐ	9°19′	1918.34	170.5±0.1	114.71±0.25	. .–11.9	2	10½

BGC	Double Star	R. A. 1880 1950	Dec. 1880 1950	Epoch	Position Angle	Distance	Magnitudes	Nights Observed	Apt.
6641	Σ 1785	13ʰ 43.6ᵐ	27°35′	1911.43	326.5±0.7	1.28±0.01	7.5– 7.9	4	10½
		46.8	14	1912.46	329.6±0.3	1.21±0.02	7.5– 7.8	4–3	10½
				1917.45	354.4±0.2	1.18±0.20	7.0– 7.7	3	10½
				1919.43	4.3±0.4	1.11±0.04	7.0– 7.3	3	10½
				1920.43	10.0±0.3	1.11±0.05	7.0– 7.2	3	10½
				1921.40	13.9±0.8	1.14±0.04	7.0– 7.2	5	10½
				1922.35	20.8±0.2	1.21±0.04	6.8– 7.1	4–3	10½
				1923.37	27.6±1.0	1.14±0.03	7.0– 7.1	4–3	10½
				1925.35	41.8±0.8	1.10±0.04	7.0– 7.2	7	10½
				1926.44	48.5±0.8	1.07±0.04	7.0– 7.3	6	10½
				1927.40	54.1	1.09	7.0– 7.6	1	10½
6668	Σ 1788	48.7	− 7 28	1888.36	75.4±0.6	2.75±0.14	6.4– 7.2	2	10
		52.4	49	1922.94	84.6±0.1	3.02±0.12	7.0– 7.9	2	10½
				1923.41	84.5	2.90	. .–. . .	1	10½
				1926.45	84.0±0.3	3.00±0.08	7.0– 7.5	3	10½
				1927.40	82.5	3.13	7.0– 7.3	1	10½
6690	β 30	52.4	20 3	1892.29	199.6±0.9	8.29±0.15	8.0–11.0	4–3	10
		55.7	19 42	1911.43	198.7±1.0	8.19±0.06	8.0–11.5	3	10½
				1916.82	198.9±0.2	8.24±0.04	8.2–11.2	2	10½
				1926.43	200.2±0.3	8.24±0.12	8.0–11.2	2	10½
6717	β 938	13 59.5	−26 0	1892.42	299.1±2.6	0.62±0.03	7.5– 7.7	4–2	10
		14 3.5	20						
6766	β 224	7.6	13 8	1916.35	47.2±3.7	0.66±0.08	8.6– 8.7	4	10½
		11.0	12 48	1917.34	47.8±3.8	0.67±0.06	8.7– 8.9	2	10½
				1918.33	44.7	0.73	. .–. . .	1	10½
6768	β 939	7.8	− 7 57	1888.31	148.4±0.8	0.70	8.0– 8.4	2–1	10
		11.5	− 8 17						
6776	Σ 1820	9.1	55 53	1893.52	72.4	2.19	8.0– 8.3	2	16
		11.4	33	1926.50	85.7	2.22	8.0– 8.3	1	10½
6780	Σ 1819	9.3	3 41	1907.41	351.4±0.8	1.09±0.07	8.2– 8.0	3	10½
		12.8	21	1911.41	349.6±0.5	1.29±0.11	8.0– 7.9	5	10½
				1913.45	346.9±0.6	1.13±0.04	8.0– 8.0	4	10½
				1914.41	346.8±1.2	1.24±0.06	8.2– 8.0	3	10½
				1917.44	343.9±0.7	1.17±0.01	8.0– 7.9	3	10½
				1919.47	341.6±0.5	1.22±0.07	7.9– 8.1	5	10½
				1922.32	338.0±0.3	1.20±0.06	8.0– 8.0	2	10½
				1926.43	333.8±1.2	1.30±0.08	8.0– 8.1	3	10½
6811	β 116	13.0	−13 9	1886.31	279.3	3.08	. .–. . .	1	26
		16.8	28	1888.37	277.4±0.4	2.98±0.05	8.2– 8.2	2	10
				1912.19	275.8±0.6	3.23±0.06	8.0– 8.4	4	10½
				1922.40	275.4±0.6	3.36±0.12	8.0– 8.4	2	10½
		14ʰ 16.8ᵐ	−13°28′	1926.44	275.6±0.7	3.22±0.02	8.0– 8.2	2	10½

BGC	Double Star	R. A. 1880 1950	Dec. 1880 1950	Epoch	Position Angle	Distance	Magnitudes	Nights Observed	Apt.
6832	Σ 1834	14ʰ 15.9ᵐ	49° 3′	1923.41	272.8±0.4	0.52±0.06	7.0– 7.4	2	10½
		18.5	48 44	1925.48	274.3±1.7	0.50±0.02	7.2– 7.3	6	10½
				1926.52	280.2±0.5	0.60±0.00	. .–. . .	2	10½
6846	β 615	17.9	49 4	1892.37	235.0±1.2	2.52±0.36	8.4– 9.8	3	10
		20.5	48 45	1926.50	238.3	2.39	8.5–10.5	1	10½
6851	Σ 1837	18.2	−11 7	1888.30	305.8±1.2	1.74±0.08	6.9– 8.2	3	10
		22.0	26	1926.44	295.6	1.38	7.0– 9.0	1	10½
6896	β 117	24.7	−15 4	1888.39	90.7±0.5	2.26±0.04	8.2– 8.9	2	10
		28.5	23	1916.61	89.9±0.8	2.22±0.06	8.2– 9.1	4	10½
6912	β 238	27.0	−20 30	1893.48	90.6	6.94	. .–. . .	1	16
		30.9	49	1917.34	88.2±0.1	7.14±0.08	8.2–10.0	2	10½
6931	β 941	29.7	0 46	1888.32	42.2±2.5	0.78±0.10	8.4– 8.5	3–2	10
		33.3	27						
. . .	AG . . .	31.6	11 33	1917.40	357.2±0.5	26.32±0.24	9.2– 9.8	5–4	10½
		35.0	15						
6955	Σ 1865 AB	35.4	14 15	1893.53	277.5	0.34	4.0– 4.1	2	16
		38.7	13 57	1911.40	141.2±0.8	0.65±0.04	3.0– 3.0	3	10½
				1912.48	142.5±0.5	0.59±0.03	4.0– 3.8	3	10½
				1913.43	142.4±0.9	0.68±0.03	4.2– 4.0	4	10½
				1914.40	140.3±0.6	0.74±0.08	4.2– 4.0	3	10½
				1917.44	140.4	0.84±0.12	4.1– 4.0	4	10½
				1919.48	140.1±0.8	0.83±0.04	4.1– 4.1	4	10½
				1919.54	137.3±1.5	0.95±0.06	4.2– 4.0	3	12
				1920.42	138.2±0.7	0.87±0.04	4.2– 4.0	4	10½
				1921.39	137.6±0.7	0.94±0.07	4.2– 4.1	4	10½
				1922.36	136.5±1.0	1.04±0.04	4.1– 4.0	4	10½
				1923.43	135.9±0.6	1.04±0.04	4.1– 4.1	6	10½
				1924.50	136.9±0.8	0.95±0.02	4.0– 4.0	4	10½
				1925.44	135.7±0.5	1.05±0.03	4.6– 4.5	5	10½
				1926.50	135.7±1.4	1.02±0.07	4.5– 4.4	11	10½
				1927.40	134.0	0.96	. .–. . .	1	10½
6955	Σ 1865 AB & C			1912.79	259.0±0.3	100.04±0.15	. .–10.5	3–2	10½
6968	β 807	36.6	− 6 18	1888.46	242.4	1.13	8.0– 9.0	1	10
		40.3	36						
6993	Σ 1877	39.8	27 35	1896.36	328.3±2.1	2.84±0.01	3–5	2	10½
		42.9	17	1926.57	332.2±1.1	2.75±0.09	. .–. . .	7	10½
6997	Σ 1876	40.0	− 6 53	1888.31	72.6±1.4	1.28±0.08	8.0– 8.0	2	10
		43.7	− 7 11	1907.45	81.0	1.25	8.0– 8.1	1	10½
		14ʰ 43.7ᵐ	− 7° 11′	1911.13	80.0±0.6	1.22±0.02	8.2– 8.4	3	10½

BGC	Double Star	R. A. 1880 1950	Dec. 1880 1950	Epoch	Position Angle	Distance	Magnitudes	Nights Observed	Apt.
6999	Σ 1879	14ʰ 40.4ᵐ 43.8	10°10′ 9 52	1922.32	111.0	1.07	8.0– 9.0	1	10½
7001	OΣ 285	41.0 43.7	42 53 35	1893.52	332.3	0.29	. .–. . .	2	16
				1923.43	86.4±1.8	0.52±0.03	7.0– 6.8	3	10½
				1925.52	83.3±2.0	0.51±0.04	7.0– 7.0	6	10½
				1926.53	84.3±2.8	0.50±0.04	7.0– 7.4	3	10½
7012	β 106	42.8 46.6	−13 39 57	1888.35	339.9±0.5	1.63±0.07	6.0– 7.2	3	10
				1907.40	343.7±0.5	1.50±0.14	6.2– 7.2	3	10½
				1916.36	345.2±0.4	1.86±0.14	6.2– 6.7	4	10½
				1919.51	349.1±0.17	1.83±0.11	6.0– 6.4	3	10½
				1926.48	349.4±0.9	1.86	6.0– 6.6	2–1	10½
7034	Σ 1888 AB	45.8 49.0	19 36 18	1892.32	240.0±1.9	3.08±0.09	. .–. . .	3	10
				1896.40	222.5±0.8	2.76±0.12	. .–. . .	3	10½
				1907.44	162.5±1.1	2.44±0.11	5.0– 7.0	3	10½
				1910.47	140.7±0.8	2.39±0.19	5.0– 7.5	3	10½
				1911.40	134.3±0.3	2.45±0.10	5.0– 7.4	4	10½
				1912.47	125.7±0.8	2.17±0.08	5.2– 7.5	4	10½
				1913.45	118.5±0.4	2.29±0.03	5.0– 7.2	3	10½
				1914.45	110.5±0.5	2.04±0.00	4.0– 7.0	2	10½
				1919.47	73.6±0.6	2.38±0.05	5.0– 7.1	6	10½
				1919.54	72.2±0.5	2.64±0.10	5.0– 7.0	3	12
				1920.49	66.1±0.8	2.52±0.08	5.0– 7.5	2	10½
				1921.44	59.7±0.2	2.56±0.03	5.0– 6.9	3	10½
				1922.42	53.4±1.1	2.81±0.13	5.0– 7.1	7	10½
				1923.40	50.3±1.8	2.92±0.07	5.0– 6.9	5	10½
				1924.50	45.0±0.9	2.94±0.06	5.0– 6.6	4	10½
				1925.49	41.7±1.2	3.13±0.13	5.0– 6.8	6	10½
				1926.46	38.2±1.0	3.19±0.06	5.0– 7.0	8	10½
7034	Σ 1888 AC			1892.24	353.8	58.98	. .–12.5	1	10
7040	β 31 AB	47.0 50.2	19 13 18 56	1916.43	197.9±1.3	1.41±0.18	8.3–10.5	4	10½
				1917.36	196.8	1.32	8.2–11.0	1	10½
				1919.51	201.8	1.45	8.0–10.5	1	10½
7040	β 31 AC			1916.42	161.6±1.1	9.10±0.62	8.5–12.5	3	10½
7040	β 31 AD			1919.51	275.7	95.11	8.0–10.8	1	10½
7041	β 118	47.0 50.9	−16 1 18	1886.31	308.7	2.00	8.0– 9.0	1	26
				1917.12	301.9±1.2	. . . . . .	8.5– 9.5	3–0	10½
7047	β 942	47.5 14ʰ 51.1ᵐ	+ 0 2 − 0°15′	1892.38	189.8±3.7	1.03±0.12	9.1– 9.2	5	10

BGC	Double Star	R. A. 1880 1950	Dec. 1880 1950	Epoch	Position Angle	Distance	Magnitudes	Nights Observed	Apt.
7049	OΣ 288	14ʰ 47.8ᵐ	16°12′	1907.40	187.8±0.4	1.65±0.07	6.5– 7.6	3	10½
		51.1	15 55	1910.47	186.4±0.6	1.55±0.05	6.5– 6.8	3	10½
				1911.41	187.2±0.4	1.58±0.07	6.4– 6.9	4–3	10½
				1912.46	187.3±0.2	1.56±0.09	6.7– 7.5	3	10½
				1914.43	187.1±0.8	1.73±0.05	7.3– 8.1	3	10½
				1917.41	186.2±0.6	1.49±0.08	6.5– 7.4	2	10½
				1918.47	186.9	1.56	6.5– 7.5	1	10½
				1919.47	186.9±0.4	1.66±0.09	6.5– 7.1	3	10½
				1922.03	185.3±0.6	1.72±0.10	6.5– 7.3	3	10½
				1923.52	185.2±0.9	1.55±0.15	6.3– 7.2	3	10½
				1924.50	186.3±0.3	1.58±0.09	6.5– 7.2	3	10½
				1925.43	186.6±0.8	1.75±0.07	6.5– 7.3	3	10½
				1926.44	184.3±1.4	1.80±0.08	6.4– 7.4	6	10½
7053	Hn 21	49.0	−14 15	1888.43	23.0	3.87	8.4– 8.8	1	10
		52.8	32	1922.93	20.7±0.9	3.82±0.18	8.8– 8.8	2	10½
7070	β 239	51.6	−27 10	1888.43	208.3	0.95	6.0– 6.0	1	10
		55.7	27	1916.45	327.3	0.83	. .–. . .	1	10½
				1917.53	327.4	0.82	. .–. . .	1	10½
				1919.55	322.9	0.97	. .–. . .	1	12
7117	β 119	14 59.2	− 6 33	1888.45	306.2±1.0	1.60±0.05	7.9– 8.4	2	10
		15 2.9	50	1892.35	307.1±0.7	1.47±0.03	8.0– 8.5	3	10
				1916.37	295.2±0.5	1.56±0.08	8.0– 8.5	2	10½
7120	Σ 1909	14 59.8	48 7	1892.41	241.6±0.3	4.54±0.01	5.0– 5.9	2	10
		15 2.1	47 50	1896.43	241.6±1.0	4.64±0.01	5.0– 6.0	2	10½
				1913.47	244.0	3.72	5.0– 5.7	1	10½
				1919.60	243.7±0.5	3.71±0.06	5.0– 6.2	3	10½
				1922.51	245.4	3.48	5.0– 6.0	1	10½
				1923.43	245.0±0.5	3.51±0.10	5.0– 5.5	3	10½
				1924.51	244.8±0.2	3.44±0.02	5.0– 5.6	2	10½
				1925.51	245.0±0.5	3.34±0.11	5.0– 5.7	4	10½
				1926.52	245.2±0.6	3.20±0.04	5.0– 5.6	6	10½
7127	Σ 1910	1.8	9 41	1888.39	211.4±0.2	4.07±0.25	6.4– 6.6	2	10
		5.2	25	1907.44	210.2±0.7	4.36±0.14	6.3– 6.5	3	10½
				1922.42	211.1±0.4	4.36±0.03	7.0– 7.4	2	10½
7132	Σ 3090	2.6	− 0 31	1885.37	276.3	1.78	. .–. . .	1	26
		6.2	47	1911.47	276.5±0.2	1.35±0.07	8.1– 8.5	3	10½
				1912.48	276.7±1.0	1.30±0.06	8.7– 8.7	3	10½
				1922.47	276.8±0.8	1.26±0.09	8.4– 8.8	2	10½
7136	β 349	2.9	2 9	1892.40	36.2±1.1	3.80±0.05	8.0–11.9	2	10
		6.4	1 53						
7137	β 809	3.0	−22 16	1892.40	119.2±0.5	1.64±0.03	8.1–10.1	2	10
		15ʰ 7.0ᵐ	−22°32′						

BGC	Double Star	R. A. 1880 1950	Dec. 1880 1950	Epoch	Position Angle	Distance	Magnitudes	Nights Observed	Apt.
. . .	Jon 441	15ʰ 3.9ᵐ 7.1	18°33′ 17	1917.41	156.2±1.9	3.70±0.44	9.7–10.9	4–3	10½
7174	β 350	8.5 12.7	−27 9 25	1892.41	160.2±0.2	1.15±0.02	7.2– 8.4	2	10
7179	Σ 1924	8.9 11.9	26 12 25 56	1896.46 1926.48	306.1±0.8 306.0±0.3	15.27±0.11 15.13±0.08	8.7– 9.1 8.8– 9.6	3 3	10½ 10½
7189	β 351	10.3 14.2	−15 8 24	1896.49	302.0±0.8	10.75±0.15	8.5–12.0	2	10½
7193	Σ 1925	10.5 14.2	− 7 50 − 8 6	1888.48 1922.48 1923.49 1925.43 1926.47	11.9 11.5±0.4 12.3 13.0±0.3 11.1	4.65 5.28±0.05 5.09 5.39±0.07 5.48	8.0– 8.6 8.0– 9.2 8.0– 9.2 8.0– 9.2 8.0– 9.2	1 3 1 3 1	10 10½ 10½ 10½ 10½
7195	β 352	10.7 14.9	−26 33 49	1892.41 1911.44 1916.98	68.3±0.1 66.2±0.5 66.2±0.4	13.95±0.01 14.18±0.09 14.04±0.11	8.1– 9.2 8.1– 9.4 8.4–10.0	2 3 2	10 10½ 10½
7201	β 227	12.1 16.2	−23 50 −24 6	1886.31 1916.41	179.0 174.5±0.8	2.33 2.13±0.15	. .−. . . 7.1–10.1	1 4	26 10½
7208	β 228	12.6 16.7	−23 50 −24 6	1886.31 1892.40 1916.40	329.2 327.6±0.4 315.0±2.8	. . . . . . 0.98±0.05 0.95±0.10	. .−. . . 7.2– 8.1 7.2– 7.7	1–0 2 2	26 10 10½
7211	Lv 6	13.: 17.2	−26 35 51	1892.38 1911.46 1916.98 1925.54	29.6±0.3 28.7±0.5 28.9±0.3 29.3	17.01±0.17 17.12±0.18 17.20±0.06 16.77	8.1– 9.5 8.2–10.2 8.2–10.8 7.7–10.0	2 3 2 1	10 10½ 10½ 10½
7214	Σ 1932	13.2 16.2	27 16 0	1893.54 1914.44 1916.52 1916.62 1917.43 1919.50 1919.55 1920.48 1921.43 1922.48 1923.40 1924.56 1925.41	320.2 360.4±2.0 363.6 369.8±2.1 9.1 11.6±0.6 15.2±1.8 14.7 18.1±0.6 19.6±0.8 22.5±1.5 26.2±0.8 27.4±0.9	0.81 0.61±0.02 0.68 0.59±0.00 0.69 0.63±0.02 0.64±0.06 0.61 0.62±0.01 0.58±0.04 0.60±0.08 0.60±0.04 0.60±0.03	6.0– 6.1 6.5– 6.7 6.2– 6.1 6.2– 6.1 6.9– 6.5 6.2– 6.0 . .−. . . 6.0– 6.2 6.0– 6.1 6.0– 5.9 6.0– 5.9 6.0– 6.2 6.0– 6.0	2 2 1 2 1 3 3 1 3 2 3 2 5	16 10½ 40 12 10½ 10½ 12 10½ 10½ 10½ 10½ 10½ 10½
		15ʰ 16.2ᵐ	27° 0′	1926.50	29.6±1.2	0.59±0.03	6.0– 6.2	6	10½

BGC	Double Star	R. A. 1880 1950	Dec. 1880 1950	Epoch	Position Angle	Distance	Magnitudes	Nights Observed	Apt.
7218	β 353	15ʰ 13.9ᵐ 12.7	85°57′ 41	1892.42	295.5	3.62	9.4– 9.8	1	10
7222	β 32	14.9 18.5	1 9 0 54	1888.44	14.6	2.79	4.5– 9.5	1	10
				1910.49	14.7±1.2	2.62±0.12	5.0– 9.2	3–2	10½
				1916.37	14.5±0.9	2.77±0.07	5.1– 9.5	4	10½
				1916.50	16.6	2.60	5.0–10.0	1	40
				1916.63	16.7	2.80	5.0–10.0	1	12
7251	Σ 1937	18.2 21.1	30 43 28	1892.48	228.5	0.66	. .–. .	1	10
				1893.50	242.8	0.50	6.0– 6.1	3	16
				1896.44	311.5±4.2	0.40±0.03	. .–. .	4	10½
				1907.40	25.6±0.2	0.98±0.09	6.1– 6.1	3	10½
				1910.47	36.2±0.4	0.90±0.08	6.0– 6.1	2	10½
				1911.47	39.4±1.1	1.00±0.02	5.8– 5.9	4	10½
				1912.49	42.2±0.1	0.97±0.05	6.0– 6.2	2	10½
				1913.45	47.0±1.2	0.89±0.04	6.0– 6.0	3	10½
				1917.45	69.0±1.3	0.70±0.01	5.0– 5.2	3	10½
				1919.53	87.0±1.1	0.65±0.03	5.6– 5.6	4	10½
				1921.42	104.7±1.3	0.58±0.01	5.0– 5.2	2	10½
				1922.48	115.2±2.0	0.54±0.02	5.0– 5.3	2	10½
				1923.40	126.1±1.4	0.50±0.02	5.0– 5.0	6	10½
				1924.54	142.6±2.8	0.53±0.02	5.0– 5.0	5	10½
				1925.47	153.6±2.4	0.55±0.02	5.0– 5.3	7	10½
				1926.45	164.6±3.4	0.64±0.03	5.0– 5.3	7	10½
7258	Σ 28 App I A & BC	20.0 22.7	37 48 33	1919.60	170.3	109.03	. .– 4.0	1	10½
7259	Σ 1938	20.0 22.7	37 46 31	1896.43	81.5±1.0	0.76±0.02	7.0– 7.4	5	10½
				1907.40	60.4±0.6	0.85±0.03	7.0– 7.2	3	10½
				1910.47	56.5±0.7	1.04±0.04	6.8– 7.1	2	10½
				1911.47	56.6±0.8	1.11±0.06	6.9– 7.1	4	10½
				1919.60	48.6±0.2	1.34±0.04	6.8– 7.2	2	10½
				1923.22	45.0±0.6	1.34±0.05	7.0– 7.2	5	10½
				1924.53	43.1±0.1	1.33±0.04	7.0– 7.3	2	10½
				1925.53	43.0±0.3	1.52±0.14	7.0– 7.4	6	10½
				1926.48	42.0±0.8	1.42±0.06	7.0– 7.4	6	10½
7293	β 33	24.7 28.5	−12 35 50	1886.41	43.0	3.16	8.2–10.5	1	26
				1892.40	42.1±0.3	3.08±0.10	7.8–10.2	2	10
				1916.46	40.0±0.8	2.68±0.07	8.0–10.7	3	10½
				1916.50	40.6	3.05	8.0–10.0	1	40
7293	β 33 AE			1916.50	312.0	28.85	8.0–12.5	1	40
7293	β 34 CD			1916.46	56.3±0.8	6.93±0.16	11.1–11.3	4–3	10½
		15ʰ 28.5ᵐ	−12°50′	1916.50	55.6	6.36	11.0–11.1	1	40

BGC	Double Star	R. A. 1880 1950	Dec. 1880 1950	Epoch	Position Angle	Distance	Magnitudes	Nights Observed	Apt.
7302	β 945	15h 26.1m 27.8	57°51′ 36	1896.51	31.8±0.1	16.03±0.08	6.6–12.0	2	10½
7318	Σ 1954	29.1 32.4	10 56 42	1926.57	182.4±0.5	3.59±0.05	3.0– 4.2	5	10½
7320	OΣ 297	29.7 32.7	25 24 10	1896.51	138.4±0.4	5.47±0.04	7.5–12.8	2	10½
7332	OΣ 298	31.8 34.3	40 12 39 58	1896.43	166.6	0.80	. .–. . .	1	10½
				1919.60	205.1	0.87	6.5– 7.0	1	10½
				1922.49	207.5	0.69	7.0– 7.4	1	10½
				1923.40	208.9±1.7	0.78±0.07	7.0– 7.3	3	10½
				1924.64	215.4±1.9	0.73±0.07	7.0– 7.2	4	10½
				1925.56	215.0±1.1	0.66±0.05	7.0– 7.2	6	10½
				1926.52	219.8±3.0	0.59±0.02	7.0– 7.2	5	10½
7336	β 121	32.3 36.5	−27 15 29	1888.50	274.2	. . . . . .	. .–. . .	1–0	10
7340	β 122	33.0 37.0	−19 23 37	1885.36	205.1	. . . . . .	. .–. . .	1–0	26
				1888.37	208.3±0.8	1.95±0.00	7.5– 7.7	3	10
				1914.44	31.8±0.4	1.93±0.01	7.0– 7.2	2	10½
				1922.42	213.1	1.81	7.0– 7.2	1	10½
7347	Howe 37	34.3 38.2	−14 26 40	1888.46	89.6±1.3	5.60±0.34	8.0– 8.5	2	10
				1911.42	88.6±0.6	5.42±0.08	9.3– 9.1	3	10½
				1922.45	89.3±2.1	5.50±0.03	8.7– 8.4	2	10½
				1926.45	89.4±0.8	5.60±0.16	9.0– 8.6	2	10½
7352	Σ 1965	34.9 37.5	37 2 36 48	1926.69	303.8±0.1	6.16	4.0– 4.9	3	10½
7359	β 35 AB	36.0 39.9	−15 38 52	1886.47	100.7	. . . . . .	8.0–11.0	1–0	26
				1916.38	100.3±0.1	2.63±0.11	7.7– 9.0	2	10½
				1922.93	101.4±0.6	2.30±0.03	7.0– 8.6	2	10½
7359	β 35 AB & C			1918.47	35.4	113.38	7.0–10.5	1	10½
7359	β 35 AC			1923.43	35.8	113.63	. .–10.0	1	10½
7367	β 619	37.6 40.9	14 3 13 49	1888.38	1.8±1.9	0.63	6.0– 7.0	2–1	10
				1907.41	3.7±0.7	0.56±0.06	6.3– 7.0	3	10½
				1923.07	4.7±1.1	0.53±0.03	6.0– 6.6	3	10½
				1924.54	359.0±1.2	0.54±0.03	6.5– 7.8	2	10½
				1925.44	359.2±1.6	0.51±0.01	6.5– 7.5	3	10½
				1926.50	355.3±2.8	0.49±0.02	6.5– 7.5	3	10½
7368	Σ 1967	37.7 15h 40.7m	26 41 26°27′	1896.43	115.1	0.48	4.0– 7.0	1	10½
				1919.54	111.2	0.69	4.0– 6.5	1	40

BGC	Double Star	R. A. 1880 1950	Dec. 1880 1950	Epoch	Position Angle	Distance	Magnitudes	Nights Observed	Apt.
7368	Σ 1967	15ʰ 40.7ᵐ	26°27′	1919.55	114.6±0.8	0.65±0.01	4.0– 6.8	2	12
				1922.45	107.3±0.8	0.60±0.01	4.0– 6.5	3	10½
				1923.44	104.4±2.5	0.47±0.05	. .–. . .	4	10½
				1925.57	104.0±4.3	0.48±0.02	4.0– 6.0	3	10½
				1926.54	98.7	0.45	4.0– 6.0	1	10½
7374	β 620 AB	38.9	−27 41	1892.41	165.5	0.65	7.0– 7.6	1	10
		43.1	55	1896.52	174.5±2.7	0.63±0.05	. .–. . .	3	10½
7374	β 620 AC			1892.14	215.4	50.53	7.0–10.0	1	10
7375	Σ 1969	39.0	60 22	1896.50	51.4±1.4	0.72±0.03	8.0– 8.3	2	10½
		40.4	8	1924.55	58.1±1.0	. . . . . .	. .–. . .	2–0	10½
7380	β 240 AB	39.5	4 24	1892.37	135.0±1.3	2.27±0.03	8.4– 9.8	3	10
		43.0	11	1911.53	135.6±1.0	2.17±0.20	8.5–10.2	3–2	10½
				1918.63	133.9	2.71	8.5–10.5	1	10½
7380	β 240 AC			1892.37	40.2	28.33	. .–12.0	2	10
				1911.53	39.1±0.5	29.16±0.06	. .–12.0	2	10½
7381	Pritchett	40.2	35 59	1911.50	223.9±0.2	5.03±0.02	9.1– 9.4	3	10½
		42.9	46	1926.48	224.5±1.4	4.66±0.00	9.2– 9.6	2	10½
7396	Σ 1974	42.9	− 2 52	1886.30	163.8	2.56	. .–. . .	1	26
		46.5	− 3 5						
7404	β 415 AB	44.8	65 57	1892.42	334.6	12.82	8.5–11.0	1	10
		45.5	44						
7404	β 415 AC			1892.42	358.5	29.12	. .–11.0	1	10
7418	β 36	46.4	−24 58	1888.58	277.8	2.60	5.8– 7.6	1	10
		50.6	−25 11	1892.39	277.8±0.5	2.86±0.02	5.8– 7.6	2	10
7421	β 810	46.9	42 50	1896.51	86.0±1.1	0.82±0.06	8.6–10.8	3	10½
		49.3	37						
7433	Σ 1985	49.7	− 1 49	1888.52	335.1	5.77	6.3– 7.4	1	10
		53.3	− 2 2	1911.44	338.2±0.3	5.85±0.03	7.0– 8.9	2	10½
				1923.51	339.2±0.6	5.56±0.09	6.8– 8.0	3	10½
				1925.57	339.1	5.68	7.0– 7.8	1	10½
				1926.51	339.2±2.4	5.70±0.03	7.0– 8.3	3	10½
7471	Σ 1992	54.6	12 1	1896.47	325.8±0.5	5.89±0.04	8.8– 8.9	3	10½
		57.9	11 49	1911.53	325.6±0.0	5.97±0.02	8.9– 9.1	2	10½
				1926.43	324.9±1.2	5.94±0.11	8.9– 8.9	3	10½
7472	β 623	54.8	− 6 38	1888.27	239.2	0.99	8.0– 9.0	1	10
		15ʰ 58.5ᵐ	− 6°50′	1914.46	234.9	1.06	8.2– 8.7	1	10½

BGC	Double Star	R. A. 1880 1950	Dec. 1880 1950	Epoch	Position Angle	Distance	Magnitudes	Nights Observed	Apt.
7476	β 37	15ʰ 55.2ᵐ 59.4	−24°15′ 27	1892.42	43.4±0.4	3.09±0.19	8.6– 9.7	3	10
7478	β 38	55.6 59.8	−24 41 53	1916.47 1917.41	351.6±1.0 351.0	4.60±0.11 4.48	7.5–10.5 8.0–11.0	2 1	10½ 10½
. . .	Jon 445	55.8 59.1	10 42 26	1917.47	293.0±0.2	2.78±0.31	10.2–10.8	2	10½
7479	H V 75	56.1 59.0	26 30 18	1896.49	111.0±0.3	47.33±0.11	7.8– 9.6	2	10½
. . .	AG Camb 7446	15 57.2 16 0.1	28 27 15	1917.49	227.6±0.2	11.56±0.13	9.0–10.2	2	10½
7480	S 676	56.5 59.2	33 40 28	1896.49	83.7±0.1	79.99±0.16	6.0– 9.2	3	10½
7487	Σ 1998 AB	15 57.8 16 1.6	−11 3 15	1888.50	21.4±0.3	1.24±0.08	5.1– 5.1	2	10
				1893.51	210.9	0.88	5.0– 5.2	2	16
				1911.44	335.5±0.2	0.58±0.05	5.0– 5.3	3	10½
				1912.49	339.6	0.66	5.0– 5.5	1	10½
				1913.47	342.9±0.8	0.69±0.03	5.0– 5.7	3	10½
				1916.53	172.1±1.2	0.93±0.06	5.0– 5.2	3	40
				1919.54	178.9±0.5	1.01±0.06	. .–. . .	3	12
				1921.52	183.4±0.1	1.13±0.04	5.0– 5.2	2	10½
				1922.45	184.6±0.6	1.15±0.05	5.0– 5.0	4	10½
				1923.43	186.1±0.6	1.15±0.05	5.0– 5.2	4	10½
				1924.59	188.0	1.14	5.0– 4.9	1	10½
				1925.54	189.4±0.8	1.13±0.05	5.0– 5.0	5	10½
				1926.55	191.0±0.6	1.11±0.06	5.0– 5.0	6	10½
7487	Σ 1998 AC			1888.54	64.0	7.98	. .– 7.8	1	10
				1893.51	67.2	7.43	. .– 7.0	2	16
7487	AB & C			1911.44	64.1±0.4	7.27±0.09	. .– 7.7	3	10½
				1913.14	64.3±0.4	7.32±0.06	. .– 8.0	3	10½
				1916.51	61.6±1.8	7.48±0.20	. .– 7.5	2	40
				1916.69	60.9	7.23	. .– 7.5	1	12
				1919.54	63.3±0.3	7.33±0.05	. .–. . .	3	12
				1922.45	60.9±1.6	7.34±0.02	. .– 7.8	3	10½
				1923.44	59.6±2.0	7.28±0.16	. .– 7.2	3	10½
				1924.59	59.8	7.02	. .– 7.5	1	10½
				1925.53	60.2±0.5	7.28±0.08	. .– 7.0	4	10½
				1926.54	59.2±1.0	7.34±0.06	. .– 9.9	4	10½
7487	Σ 1998 BC			1888.50	69.3±0.2	7.12±0.14	. .– 7.5	2	10
7488	Σ 1999	15 57.8 16ʰ 1.6ᵐ	−11 7 −11°19′	1913.48	99.9	11.15	8.0– 8.7	1	10½

BGC	Double Star	R. A. 1880 / 1950	Dec. 1880 / 1950	Epoch	Position Angle	Distance	Magnitudes	Nights Observed	Apt.
. . .	Jon 446	15ʰ 57.8ᵐ / 16 1.1	10°41′ / 29	1917.47	3.4±0.6	4.04±0.17	9.8–10.0	2	10½
7495	β 948 AB	15 59.3 / 16 3.0	− 5 58 / − 6 10	1888.46 / 1892.40	148.2±0.4 / 149.3±1.1	1.54±0.01 / 1.54±0.09	6.9– 9.4 / 7.1– 9.4	2 / 4–3	10 / 10
7495	β 948 AC			1892.41	234.8±0.2	29.05±0.10	. .–10.4	2	10
7495	β 948 AD			1892.41	194.9±0.1	52.72±0.06	. .–10.8	2	10
7502	β 39	1.0 / 4.9	−12 25 / 37	1888.44 / 1923.44	260.3 / 253.4	3.30 / 3.28	5.5–10.0 / 6.0–10.5	1 / 1	10 / 10½
7506	β 949	1.9 / 5.7	− 9 47 / 59	1916.61	195.3±1.1	0.20±0.03	7.5– 8.0	2	40
7533	β 120 AB	5.0 / 9.1	−19 9 / 20	1886.30	7.2	0.65	. .–. . .	1	26
				1888.40	4.6±1.3	0.81±0.00	4.0– 5.0	3	10
				1910.49	2.2±1.0	0.82±0.01	4.3– 5.7	3	10½
				1917.48	3.0±1.3	0.80±0.07	4.0– 5.0	2	10½
				1919.56	1.9±0.7	0.93±0.08	. .–. . .	4	12
				1921.54	4.3±2.0	0.94±0.11	4.0– 5.5	3	10½
				1923.54	5.3	1.02	. .–. . .	1	10½
7533	β 120 AB & C			1910.47	336.2±0.0	41.36±0.08	. .–. . .	2	10½
				1917.48	336.7±0.6	41.27±0.16	4.0– 7.0	3	10½
				1919.56	336.4±0.4	41.13±0.07	. .–. . .	2	12
7533	Mitchell CD			1885.38	44.2	. . . . . .	. .–. . .	1–0	26
				1888.40	49.7±0.7	2.24±0.14	7.4– 8.1	3	10
				1910.49	49.6±1.9	2.18±0.04	7.4– 8.1	3	10½
				1916.86	50.7±1.3	2.19±0.08	7.6– 8.5	5–4	10½
				1919.56	50.5±1.7	2.29±0.03	7.5– 7.8	4	12
				1921.54	50.3±1.0	2.25±0.10	7.5– 8.2	3	10½
7561	Σ 2026	10.1 / 13.5	7 41 / 30	1896.51 / 1923.41	263.0±2.0 / 59.7±1.5	0.63±0.00 / 0.83±0.11	8.7– 9.2 / 8.5– 9.0	3 / 2	10½ / 10½
7563	Σ 2032 AB	10.2 / 12.8	34 10 / 33 59	1893.49	210.4	4.09	6.0– 7.5	1	16
				1896.44	211.1±0.5	4.30±0.14	5.0– 6.0	4	10½
				1910.53	215.5±0.6	4.78±0.05	5.0– 6.2	3	10½
				1911.51	216.0±0.5	4.84±0.04	5.0– 6.0	4–3	10½
				1918.65	219.2±0.4	5.11±0.08	5.0– 6.0	4	10½
				1919.61	219.0±0.7	5.11±0.09	5.0– 6.4	4	10½
				1921.64	220.0±0.4	5.11±0.05	5.0– 6.3	3	10½
				1922.76	219.5±1.1	5.16±0.03	5.0– 6.1	3	10½
				1923.51	220.5±0.1	5.14±0.00	5.0– 5.8	2	10½
				1924.56	220.0±0.3	5.18±0.04	5.0– 6.1	3	10½
				1925.56	220.4±0.3	5.14±0.05	5.0– 6.0	6	10½
		16ʰ 12.8ᵐ	33°59′	1926.51	220.2±1.2	5.20±0.06	5.0 –6.2	5	10½

BGC	Double Star	R. A. 1880 1950	Dec. 1880 1950	Epoch	Position Angle	Distance	Magnitudes	Nights Observed	Apt.
7563	Σ 2032 AD	16ʰ 12.8ᵐ	33°59′	1910.53	85.9±0.3	64.40±0.25	. .—. . .	3	10½
				1911.50	85.8±0.1	64.95±0.39	. .– 9.5	3	10½
				1919.61	84.9±0.4	67.05±0.21	. .– 9.5	3	10½
				1926.49	85.0	68.61	. .—. . .	2	10½
7565	Σ 2036 AB	10.5	72 52	1896.52	233.0±0.7	2.33±0.13	9.0–10.1	4	10½
		7.0	41						
7565	Σ 2036 AC			1896.56	340.6±1.7	14.27±0.08	. .–13.0	3–2	10½
7589	Σ 2041	15.6	1 31	1888.50	1.3	2.46	7.8–10.0	1	10
		19.1	21						
7613	Sh 228 AB	18.4	−23 10	1886.51	357.3	3.76	7.8– 8.2	1	26
		22.6	20						
7617	β 950	18.7	− 9 35	1888.46	351.5	. . . . . .	8.0–10.0	1–0	10
		22.5	45	1911.99	353.4±2.0	0.84±0.04	8.0– 9.5	2	10½
7631	Mh	22.0	−26 10	1888.55	275.5±0.5	3.30±0.13	1.0– 6.5	3	10
		26.3	20	1926.57	275.7±1.6	3.35±0.11	1.2– 6.8	4	10½
7636	OΣ 311	22.6	21 10	1896.46	198.5±0.4	7.34±0.13	7.7–10.8	3	10½
		25.6	0						
7640	β 815	23.3	43 11	1896.47	341.6±0.5	8.65±0.16	8.1–10.3	3	10½
		25.5	1						
7642	Σ 2052	23.6	18 40	1925.54	268.9±0.7	0.77±0.05	7.7– 7.8	6	10½
		26.7	31	1926.54	266.4±0.4	0.73±0.01	7.9– 7.6	6	10½
7649	Σ 2055	24.9	2 15	1888.39	44.5±1.7	1.52±0.09	4.1– 5.9	3–4	10
		28.4	6	1893.53	48.8	1.43	4.0– 5.2	2	16
				1909.46	64.7±1.3	1.18±0.10	4.7– 6.3	3	10½
				1911.48	67.9±0.6	1.18±0.06	4.8– 5.6	4	10½
				1913.53	70.9±0.4	0.98±0.04	4.0– 5.6	5	10½
				1917.47	78.2±1.4	0.88±0.08	4.0– 6.0	2	10½
				1918.64	78.2±1.0	0.87±0.04	4.0– 6.1	4	10½
				1919.54	77.4±0.0	0.91±0.01	4.0– 5.5	2	12
				1919.55	79.7±0.1	1.06±0.00	4.0– 5.0	2	40
				1919.56	79.3±0.6	0.87±0.04	4.0– 6.1	3	10½
				1921.54	84.5±1.0	0.82±0.08	4.0– 5.1	3	10½
				1922.46	85.0±1.0	0.85±0.04	4.0– 5.2	4	10½
				1923.42	86.6±0.6	0.80±0.04	4.0– 4.8	4	10½
				1924.59	90.6±1.3	0.73±0.01	4.0– 5.0	4	10½
				1925.42	93.3±0.8	0.73±0.04	4.0– 4.8	6	10½
				1926.54	98.0±0.6	0.70±0.02	4.0– 5.8	8	10½
7651	Σ 3105	25.4	− 6 46	1886.30	41.5	0.51	. .—. . .	1	26
		16ʰ 29.1ᵐ	− 6°55′						

BGC	Double Star	R. A. 1880 1950	Dec. 1880 1950	Epoch	Position Angle	Distance	Magnitudes	Nights Observed	Apt.
7661	Σ 2057	16ʰ 26.3ᵐ	19°33′	1896.48	267.2±0.2	4.94±0.02	9.3– 9.3	2	10½
		29.4	24	1922.48	267.6±1.0	5.06±0.02	9.0– 8.8	2	10½
				1926.51	265.7±0.4	5.27±0.13	9.0– 9.1	3	10½
7663	Σ 2058 AB	26.4	19 34	1896.49	347.6±0.4	1.96±0.07	9.2– 9.4	2	10½
		29.5	25						
7663	Σ 2058 AB & C			1896.49	137.7±0.6	15.78±0.17	. .–10.5	2	10½
7669	β 817	27.5	23 29	1909.46	326.0±1.5	1.11±0.09	8.0– 8.0	3	10½
		30.5	20	1922.50	325.9±1.2	1.10±0.02	8.2– 8.4	3	10½
				1923.51	324.4±1.1	1.14±0.16	8.2– 8.3	2	10½
7671	Σ 2075	28.1	80 19	1896.60	321.0±0.8	1.15±0.05	8.2–11.4	2	10½
		26.0	10						
7692	Hn 26	31.2	− 5 16	1888.52	183.2	6.80	9.0– 9.2	1	10
		34.9	25	1922.50	184.8±0.8	6.80±0.28	9.3– 9.4	2	10½
7708	Σ 2080	34.4	38 34	1896.54	25.9±0.4	3.26±0.12	8.0–11.8	4	10½
		36.8	25						
7712	β 42	35.3	29 15	1916.49	38.2	7.31	9.2– 9.4	1	10½
		38.1	7	1917.43	39.2±0.5	7.54±0.16	9.0– 9.3	5	10½
				1926.58	41.2±0.8	7.57±0.09	9.0– 9.1	3	10½
7717	Σ 2084	36.8	31 49	1906.81	176.5±0.5	1.13±0.04	. .–. . .	2	10½
		39.5	41	1911.56	137.0±2.7	1.22±0.04	3.0– 6.6	6–5	10½
				1913.80	121.4±2.2	1.17±0.08	3.0– 6.5	4	10½
				1917.48	101.0±1.9	1.64±0.12	3.0– 6.2	2	10½
				1918.70	96.8±1.3	1.73±0.10	3.0– 6.0	7	10½
				1919.61	89.8±0.6	1.74±0.04	3.0– 6.0	2	12
				1919.61	89.0±1.0	1.73±0.08	3.0– 6.2	6–5	10½
				1921.64	80.6±0.4	1.61±0.08	3.0– 6.5	3	10½
				1922.73	75.0±0.9	1.69±0.09	3.0– 6.2	4	10½
				1923.60	70.6±0.8	1.66±0.11	3.0– 6.5	3	10½
				1924.54	65.1±1.5	1.76±0.13	3.0– 6.0	3	10½
				1925.58	60.3±1.2	1.60±0.13	3.0– 6.2	6	10½
				1926.70	52.9±0.8	1.36±0.10	3.0– 6.0	6	10½
7748	Δ 15	40.2	43 42	1893.54	347.6	0.44	8.0– 8.4	2	16
		42.4	34	1896.46	336.8±1.2	0.58±0.02	8.0– 8.3	4–2	10½
				1911.44	278.4±2.4	0.46±0.00	8.0– 8.4	2	10½
				1919.62	238.1	. . . . . .	8.0– 8.2	1–0	10½
				1923.45	212.8±3.6	0.52±0.02	8.0– 8.1	3	10½
				1925.58	205.8±1.6	0.61±0.00	8.0– 8.1	2	10½
				1926.57	202.1±3.6	0.53±0.02	8.0– 8.3	2	10½
7763	β 43	42.3	2 57	1916.47	241.8±0.7	1.22±0.07	8.5– 8.6	3	10½
		16ʰ 45.8ᵐ	2°49′	1916.53	242.6±0.7	1.03	9.0– 9.2	2–1	40

BGC	Double Star	R. A. 1880 1950	Dec. 1880 1950	Epoch	Position Angle	Distance	Magnitudes	Nights Observed	Apt.
7763	β 43	16ʰ 45.8ᵐ	2°49′	1917.42	242.5±0.5	1.12±0.02	8.7– 8.5	2	10½
				1926.58	59.6	1.20	8.5– 8.6	1	10½
7777	OΣ 315	45.3	1 25	1888.51	163.0±0.9	1.01±0.11	6.1– 8.0	3	10
		48.8	18	1923.45	150.0±0.7	0.79±0.04	6.0– 7.7	3	10½
				1924.55	148.5±0.8	0.78±0.06	6.0– 7.5	3	10½
				1925.55	147.2±3.8	0.84±0.06	6.0– 7.5	2	10½
				1926.58	147.7	0.79	6.0– 8.0	1	10½
7778	Σ 2106	45.4	9 37	1916.59	283.4±2.0	0.25±0.02	6.8– 8.2	3	40
		48.7	30						
7779	β 627	45.7	46 12	1896.49	317.3±1.2	1.82±0.22	5.0– 9.5	5–4	10½
		47.7	5						
7783	Σ 2107	47.1	28 52	1896.52	304.1±0.6	0.35±0.04	6.0– 8.0	3	10½
		49.9	45	1911.50	12.4±0.6	0.53±0.01	6.0– 7.8	2	10½
				1918.70	29.7±1.8	0.63±0.05	6.1– 7.6	5	10½
				1919.60	29.6±2.3	0.66±0.01	6.5– 8.5	4	10½
				1923.45	35.5±1.4	0.67±0.07	6.5– 7.6	4	10½
				1924.66	38.9±1.2	0.75±0.04	6.5– 7.7	5	10½
				1925.57	40.8±1.1	0.71±0.06	6.5– 7.5	4	10½
				1926.57	39.7±2.3	0.77±0.07	6.5– 8.2	3	10½
7786	β 123	47.5	−21 51	1892.48	203.9±1.8	1.57±0.02	8.4– 8.8	3	10
		51.7	58	1916.47	196.6	1.35	8.5– 8.8	1	10½
7791	β 241	48.4	−21 22	1888.56	165.0±2.2	. . . . . .	7.2– 7.2	2–0	10
		52.6	29	1916.47	173.0	0.52±0.00	7.0– 7.0	1–2	10½
				1917.47	168.7±0.2	0.61±0.02	. .–. . .	2	10½
7792	Ku 1	48.4	77 43	1896.51	192.7±0.2	2.90±0.03	6.6–10.2	3–2	10½
		47.2	36	1926.48	183.4	2.99	7.0–11.0	1	10½
7798	Σ 3106	49.3	− 4 59	1886.41	67.2	2.33	8.5– 8.7	1	26
		53.0	− 5 6	1925.57	250.2±0.6	2.11±0.10	8.5– 8.6	2	10½
7801	β 1117	49.6	−22 57	1893.50	270.6	0.56	6.1 –6.4	2	16
		53.8	−23 4						
7803	Sh 240	50.0	−19 21	1888.52	232.9	4.78	6.0– 7.3	1	10
		54.1	28	1911.56	231.8±0.8	4.93±0.13	6.5– 7.8	2	10½
7813	Hn 27	52.5	−13 1	1888.50	134.2±1.6	5.06	8.4– 9.5	2–1	10
		56.4	8						
7817	Σ 3107	52.9	4 9	1888.51	96.3±1.0	1.27±0.14	8.2– 8.6	3	10
		56.4	2	1893.52	97.0	1.32	8.0– 8.3	2	16
		16ʰ 56.4ᵐ	4° 2′	1924.50	90.7	1.78	8.0– 8.4	1	10½

BGC	Double Star	R. A. 1880 1950	Dec. 1880 1950	Epoch	Position Angle	Distance	Magnitudes	Nights Observed	Apt.
7831	Σ 2117	16ʰ 55.2ᵐ	51°59′	1896.58	105.8±0.4	1.32±0.06	8.2–10.8	2	10½
		56.2	52						
7834	Σ 2118	55.8	65 13	1921.76	78.2±0.4	0.58±0.01	. .–. . .	2	10½
		56.2	7	1923.53	77.2±1.1	0.63±0.04	6.7– 6.8	6	10½
				1924.65	76.6±2.8	0.64±0.04	6.5– 6.9	5	10½
				1925.60	77.7±1.2	0.63±0.01	6.5– 7.0	4	10½
7847	β 822	16 58.7	19 51	1896.52	228.1±2.9	1.64±0.07	6.8–11.2	4–3	10½
		17 1.7	5						
7856	β 357	16 59.9	10 43	1896.51	298.0±0.6	1.20±0.03	8.0– 9.3	3	10½
		17 3.2	37	1911.45	298.0±0.6	1.14±0.10	8.0– 9.4	3	10½
7858	Σ 2120	0.0	28 15	1926.55	237.4±0.2	11.44±0.10	6.5– 9.1	6	10½
		2.8	9						
7863	β 823	0.5	— 0 49	1888.52	358.8	. . . . . .	8.0– 9.3	1–0	10
		4.1	55	1896.49	5.3±0.3	0.95±0.02	8.0– 9.2	3	10½
				1911.45	19.3±0.6	0.89±0.05	8.0– 9.2	3	10½
7878	Σ 2130	2.8	54 38	1893.53	155.2	2.32	5.0– 5.1	2	16
		4.3	32	1921.69	122.8±0.0	2.25±0.09	5.1– 5.1	3	10½
				1922.75	121.5±0.9	2.23±0.09	5.0– 5.1	4	10½
				1923.53	120.8±1.1	2.22±0.06	5.0– 5.0	5	10½
				1924.56	120.5±1.0	2.27±0.10	5.0– 5.2	3	10½
				1925.60	118.1±0.2	2.13±0.05	5.0– 5.1	3	10½
				1926.77	116.7±0.7	2.37±0.03	5.0– 5.2	3	10½
7885	β 1118 AB	3.5	—15 34	1896.52	259.8±1.6	0.38±0.01	3.0– 4.0	3	10½
		7.5	40	1910.54	244.2±1.1	0.55±0.15	3.0– 3.4	4–3	10½
				1911.54	242.2±0.2	0.56±0.06	3.5– 4.2	2	10½
				1916.57	241.8	0.51	3.0– 3.6	1	12
				1916.59	240.6±2.0	0.59±0.06	3.0– 3.6	2	40
				1921.68	234.9±0.6	0.62±0.01	4.0– 5.0	2	10½
				1922.57	235.7±0.8	0.66±0.02	3.0– 3.5	3	10½
				1923.53	232.9±0.2	0.67±0.03	3.0– 3.8	3	10½
				1925.58	233.6±0.5	0.65±0.02	3.0– 3.8	3	10½
				1926.55	233.3±0.9	0.64±0.00	3.0– 3.8	3	10½
7885	β 1118 AB & D			1911.03	287.8±0.2	100.50±0.32	. .–11.5	2	10½
7887	β 124	4.0	— 0 36	1916.60	263.1±2.4	0.98±0.03	7.3–10.0	2	40
		7.6	42	1916.63	266.0	1.12	7.3–10.0	1	12
				1917.44	269.7	0.99	7.5–10.5	1	10½
7891	β 125	4.7	—26 53	1919.55	65.7	1.93	8.0–11.0	1	12
		17ʰ 9.0ᵐ	—26°59′						

BGC	Double Star	R. A. 1880 1950	Dec. 1880 1950	Epoch	Position Angle	Distance	Magnitudes	Nights Observed	Apt.
7900	OΣ 325	17^h 7.2^m 10.6	7°54' 49	1888.57	206.6	1.35	7.5– 9.0	1	10
				1925.60	212.0	1.08	7.5– 9.5	1	10½
				1926.58	213.0	1.16	7.5–10.2	1	10½
7905	Sh 243	8.0 12.3	−26 25 30	1888.39	197.5±0.9	4.33±0.13	4.2– 5.3	3	10
				1919.62	183.0	4.25	. .–. . .	1	10½
				1922.50	182.4	4.28	5.5– 5.6	1	10½
				1923.51	180.6	4.36	5.5– 5.6	1	10½
				1925.59	179.6±0.5	4.12±0.10	5.6– 5.7	2	10½
7912	β 957	9.0 12.9	−10 10 15	1888.54	20.4	. . . . . .	8.0– 8.0	1	10
7914	Σ 2140	9.2 12.4	14 32 27	1919.65	112.7±0.4	4.57±0.07	3.0– 6.2	3	10½
				1921.64	111.7±0.4	4.64±0.06	3.5– 5.5	3	10½
				1922.73	111.6±0.3	4.57±0.04	3.5– 5.8	3	10½
				1923.56	112.3±0.5	4.56±0.13	3.5– 6.5	3	10½
				1924.61	110.8±0.5	4.64±0.01	3.0– 6.0	2	10½
				1925.56	111.1±0.6	4.54±0.02	. .–. . .	2	10½
				1926.50	111.0±0.6	4.43±0.09	3.0– 6.0	3	10½
7915	β 44 AB	9.2 11.9	28 57 52	1892.50	18.7±0.6	5.44	8.8– 9.5	2–1	10
				1910.89	17.3±0.3	5.54±0.08	8.3– 9.2	3–2	10½
				1916.46	15.3±0.4	5.53±0.10	8.6–10.3	4	10½
				1925.51	17.0±1.1	5.55±0.12	8.5– 9.8	3	10½
				1926.60	16.0±1.1	5.54±0.24	8.5– 9.6	4	10½
7915	β 44 AC			1916.48	43.0±0.2	39.02±0.10	. .–11.1	2	10½
				1925.51	44.1±0.3	39.98±0.17	. .–10.7	3	10½
				1926.53	44.9	40.82	. .–. . .	1	10½
7917	β 958	9.4 13.5	−19 12 17	1888.45	217.6	1.36	8.3– 9.2	1	10
7922	Σ 3127	10.1 13.0	24 59 54	1889.62	186.7±0.4	16.20±0.06	. .–. . .	39–11	10
				1890.79	187.2±0.6	16.04	. .–. . .	6–1	10
				1891.43	187.4±0.6	15.95±0.09	. .–. . .	83–24	10
				1922.77	203.7±0.3	11.51±0.08	3.0– 7.5	4	10½
				1923.51	204.3±0.4	11.42±0.25	3.0– 7.5	2	10½
				1926.52	205.6±0.0	11.19±0.01	. .–. . .	2	10½
7923	S 685	10.2 14.5	−26 30 35	1888.49	335.8±0.4	5.77±0.09	6.0– 8.0	2	10
				1923.51	335.3	5.58	6.0– 8.0	1	10½
				1925.59	336.4	6.05	6.0– 8.0	1	10½
7935	H. C. Wilson 15	11.8 17^h 14.6^m	26 43 26°38'	1893.49	40.8	0.60	8.0– 9.5	1	16

BGC	Double Star	R. A. 1880 1950	Dec. 1880 1950	Epoch	Position Angle	Distance	Magnitudes	Nights Observed	Apt.
7936	OΣ 327	17^h 11.9^m	56°16′	1896.58	316.0±2.5	0.30±0.02	7.5– 7.7	2	10½
		13.2	11	1921.68	151.0±3.0	0.42±0.02	. .–. . .	2	10½
				1923.53	148.9±0.5	0.45±0.03	7.5– 7.8	2	10½
				1925.58	334.3	0.42	7.5– 7.9	1	10½
7943	β 126 AB	12.9	−17 38	1888.54	264.2±0.7	1.86±0.02	6.2– 8.3	2	10
		17.0	43	1911.53	262.7±0.9	1.87±0.07	6.0– 7.3	3	10½
				1916.46	262.6±1.4	1.94±0.12	6.0– 8.0	3–2	10½
				1919.63	263.2±0.6	1.96±0.06	6.0– 6.8	2	10½
7943	β 126 AC			1911.53	140.10	11.40	. .–11.5	1	10½
				1916.46	141.2±0.0	11.65±0.17	. .–11.6	2	10½
				1919.63	138.0	11.31	. .–11.3	1	10½
7952	β 45	13.5	32 37	1916.45	290.3±0.8	4.93±0.11	8.7– 9.2	2	10½
		16.1	32						
7968	Σ 2150 AB	15.2	1 41	1896.48	196.4±0.5	9.22±0.12	9.3– 9.7	3	10½
		18.7	36	1925.53	202.8±0.1	9.74±0.16	9.3– 9.6	4	10½
				1926.51	202.0±0.6	9.72±0.16	9.3– 9.7	4	10½
7968	Σ 2150 AC			1925.53	57.2±0.1	163.92±0.26	. .– 7.6	3	10½
				1926.48	57.1±0.1	164.64±0.00	. .– 7.7	2	10½
7984	β 242 AB	17.4	−11 35	1888.50	75.1	0.97	8.0– 9.3	1	10
		21.3	39	1892.50	72.4±0.4	0.96±0.04	8.2– 8.7	2	10
				1916.46	71.0±3.3	1.00±0.02	8.0– 9.5	2	10½
7984	β 242 AC			1888.50	66.1	9.64	8.0–11.2	1	10
				1892.48	62.0	9.04	. .–11.0	1	10
7984	AB & C			1916.47	62.5±0.1	9.04±0.30	. .–11.3	3	10½
7984	β 242 AB & D			1892.48	64.8	47.83	. .–10.0	1	10
				1916.47	64.2±0.5	47.90±0.02	. .–10.5	2	10½
7994	β 46	18.1	13 31	1892.51	202.9±1.4	2.00±0.15	8.0–11.2	3–2	10
		21.3	27	1916.51	204.6±0.4	2.02±0.03	8.0–10.9	2	40
				1917.43	203.3±0.1	1.91±0.10	7.6–11.1	2	10½
8003	Σ 2161	19.6	37 15	1926.76	314.7±0.5	3.72±0.05	4.0– 5.1	3	10½
		22.0	11						
8014	β 129	21.2	−25 24	1916.69	105.0	0.85	8.0– 8.3	1	12
		25.5	28	1917.48	100.3±0.4	1.00±0.10	7.5– 8.1	2	10½
8025	Σ 2171	22.7	− 9 54	1888.46	69.4±1.0	1.45±0.05	8.0– 8.2	3–2	10
		17^h 25.4^m	− 9°58′	1911.57	68.8±1.6	1.40±0.10	8.3– 8.2	3	10½

BGC	Double Star	R. A. 1880 1950	Dec. 1880 1950	Epoch	Position Angle	Distance	Magnitudes	Nights Observed	Apt.
8038	Σ 2173	17ʰ 24.2ᵐ 27.8	— 0°58′ — 1 2	1888.49	168.7±0.6	0.68±0.03	6.0– 6.0	3	10
				1893.55	157.4	1.14	. .–. . .	1	16
				1916.57	159.6	0.59	6.0– 6.4	1	12
				1916.60	158.4	0.65	6.0– 6.4	1	40
				1918.63	155.9±0.10	0.84±0.03	6.0– 6.4	4	10½
				1919.54	151.8±1.8	0.89±0.10	6.0– 6.2	3	12
				1921.61	149.5±1.2	0.75±0.04	6.1– 6.1	3	10½
				1922.48	145.1±2.6	0.78±0.06	6.3– 6.2	3	10½
				1923.43	143.9±2.2	0.72±0.04	6.1– 6.2	4	10½
				1924.54	140.1±2.3	0.62±0.02	6.2– 6.0	4	10½
				1925.57	133.2±1.4	0.56±0.02	6.3– 6.1	4	10½
				1926.58	131.8	0.52	. .–. . .	1	10½
8057	Σ 2180	26.1 27.8	50 58 55	1921.74	262.6±0.9	3.06±0.00	7.0– 7.2	2	10½
				1926.58	261.5	3.00	7.0– 7.1	1	10½
8080	Hn 30	30.5 34.7	—23 19 22	1888.54	104.3	3.59	8.7–10.5	1	10
				1925.59	105.0±2.2	3.34±0.04	8.0– 9.8	2	10½
8099	β 962	33.8 34.5	61 58 55	1921.72	334.2±1.2	1.81±0.27	5.5–10.0	2	10½
				1922.73	331.8	1.77	5.5– 9.5	1	10½
				1923.53	331.3±1.2	1.75±0.03	5.5–10.0	3	10½
				1924.56	327.9±1.9	1.81	5.5–10.3	3–1	10½
8100	β 631	33.8 37.4	— 0 35 38	1888.51	58.8±4.2	0.42	7.0– 7.0	3–1	10
				1911.22	52.2±1.6	0.42±0.04	6.3– 6.0	3	10½
				1916.52	57.0±1.1	0.43±0.01	7.4– 7.2	2	40
				1925.57	42.2±1.6	0.42±0.02	7.5– 7.0	2	10½
				1926.58	35.4	0.41	. .–. . .	1	10½
8133	A 233	38.3 41.2	24 51 49	1914.62	237.9±0.1	3.28±0.12	8.0–12.8	2	40
8162	A. Clark 7 BC	41.8 43.6	27 48 46	1914.60	154.8±1.3	0.44±0.03	10.0–10.7	3	40
				1919.56	224.1	0.89	10.0–10.5	1	40
				1922.48	241.4±2.3	1.04	9.5–10.5	2	10½
8162	A. Clark 7 A & BC			1914.60	245.7±0.2	32.51±0.15	. .–. . .	3	40
				1919.60	245.7±0.3	33.17±0.14	. .–. . .	4	10½
8209	Σ 3128	46.5 50.3	— 7 53 54	1886.30	26.0	1.19	7.7–11.0	1	26
8210	OΣ 338	46.6 49.8	15 21 20	1888.50	20.7±1.4	0.75±0.11	7.0– 6.2	3	10
				1911.52	11.3±0.3	0.78±0.03	6.8– 6.5	3	10½
				1921.60	8.2±0.3	0.72±0.02	7.0– 7.0	2	10½
				1922.50	7.7±0.5	0.76±0.00	7.1– 7.0	2	10½
				1923.49	8.2±1.4	0.82±0.10	7.3– 7.0	2	10½
				1924.66	8.1±1.6	0.74±0.04	7.1– 6.8	3	10½
				1925.55	7.0±0.4	0.77±0.00	6.8– 6.9	3	10½
		17ʰ 49.8ᵐ	15°20′	1926.64	7.2±0.5	0.75±0.02	7.0– 6.8	3	10½

BGC	Double Star	R. A. 1880 1950	Dec. 1880 1950	Epoch	Position Angle	Distance	Magnitudes	Nights Observed	Apt.
8235	β 130	17ʰ 49.4ᵐ	40° 2′	1914.57	121.4±1.5	1.88±0.04	5.7– 9.2	3	40
		51.7	1	1916.77	122.7±1.5	1.75±0.24	6.0– 9.7	4–3	10½
8241	Σ 2244	50.9	0 5	1888.58	274.4±1.3	1.00±0.00	6.5– 6.5	2	10
		54.5	4	1910.53	276.0±0.1	0.85±0.01	6.8– 7.2	2	10½
8288	β 47	54.9	−10 14	1896.55	273.2±0.6	1.64±0.09	8.1–10.8	3	10½
		58.8	15	1916.67	280.9	1.38	8.7–10.5	1	40
				1917.49	277.5±1.2	1.20±0.08	8.5–10.7	3–2	10½
8303	Σ 2262	17 56.6	− 8 11	1888.51	255.7±0.3	1.91±0.08	5.4– 6.4	3	10
		18 0.4	11	1911.64	261.2±0.2	2.10±0.06	5.0– 6.0	2	10½
				1913.57	260.8±0.5	2.00±0.04	6.0– 7.0	3	10½
				1914.60	260.9±1.0	2.15±0.08	5.0– 5.6	4	40
				1915.70	262.0±0.5	1.92±0.09	5.9– 6.8	4	10½
				1918.65	262.2±0.4	2.00±0.06	5.0– 5.6	4	10½
				1919.65	261.8±0.7	1.98±0.3	5.5– 6.0	2	10½
				1919.68	262.0	2.00	5.0– 5.4	1	12
				1921.63	261.0±0.7	1.99±0.06	5.3– 5.7	3	10½
				1922.58	262.1±0.6	2.05±0.07	5.0– 5.9	3	10½
				1923.54	262.4±0.7	2.08±0.09	5.0– 5.6	5	10½
				1924.68	263.3±0.6	2.13±0.08	5.0– 5.9	4	10½
				1925.59	263.4±0.3	2.11±0.04	5.0– 5.7	5	10½
				1926.58	263.0±0.2	2.01±0.04	5.0– 5.4	5	10½
8340	Σ 2272	17 59.4	2 33	1888.65	353.0	2.15	4.5– 6.5	1	10
		18 2.9	33	1896.73	288.6±0.5	2.19±0.02	4.0– 5.5	6	10½
				1906.82	170.2±0.8	2.67±0.15	. .–. . .	6	10½
				1909.70	156.5±0.3	3.21±0.06	. .–. . .	3	10½
				1910.53	152.0±0.1	3.54±0.04	4.0– 5.5	3	10½
				1911.63	149.0±0.4	3.78±0.05	4.0– 5.7	4	10½
				1914.63	142.6±0.6	4.36±0.09	4.0– 6.2	3	40
				1915.68	139.8±0.2	4.58±0.04	4.0– 6.2	3	10½
				1918.67	135.6±0.5	5.26±0.13	4.0– 6.2	6	10½
				1919.68	134.5	5.38	4.0– 6.3	1	12
				1919.74	134.2±0.1	5.28±0.04	4.0– 5.9	4	10½
				1920.79	133.2±0.7	5.41±0.08	4.0– 5.7	4	10½
				1921.63	131.7±0.4	5.57±0.04	4.5– 6.0	4	10½
				1922.73	130.4±0.1	5.79±0.06	4.5– 6.2	4	10½
				1923.58	129.4±0.6	5.82±0.10	4.5– 6.2	5	10½
				1924.58	128.4±0.1	5.84±0.05	4.5– 5.6	6	10½
				1925.61	127.1±0.5	6.10±0.13	4.5– 5.7	4	10½
				1926.63	126.5±0.3	6.16±0.08	4.5– 6.0	7	10½
8340	Σ 2272 Aa			1914.63	223.4±0.2	40.60±0.14	. .–12.0	2	40
8355	β 243 AB	0.9	−22 17	1916.67	122.5±0.0	0.66	8.0– 8.5	2–1	12
		18ʰ 5.1ᵐ	−22°17′	1917.50	126.6	0.89	. .–. . .	1	10½

BGC	Double Star	R. A. 1880 1950	Dec. 1880 1950	Epoch	Position Angle	Distance	Magnitudes	Nights Observed	Apt.
8356	β 244	18ʰ 1.0ᵐ 5.4	−27°53′ 53	1892.54	257.9±0.3	2.04±0.06	8.0–10.3	2	10
				1916.69	260.0	2.50	8.0– 9.3	1	12
				1917.47	262.0±2.8	1.96±0.12	8.0– 9.2	2	10½
8367	β 636 AB	2.1 5.6	2 12 12	1914.75	126.5	4.34	7.5–13.0	1	40
8367	β 636 AC			1914.75	103.6	14.34	. .–14.5	1	40
8368	β 826	2.1 5.4	9 45 45	1893.52	333.2	0.59	9.8– 9.9	2	16
8370	OΣ 524	2.3 5.3	19 39 39	1916.59	37.9±0.7	0.20±0.00	7.0– 8.2	2	40
8371	β 245	2.4 6.9	−30 45 45	1888.54	362.5	4.12	6.0– 8.5	1	10
				1892.52	353.3±0.5	3.98±0.01	6.4– 8.9	2	10
				1916.63	352.6±0.6	3.72±0.04	6.0– 8.8	2	12
				1916.77	353.1	4.02	6.0– 8.5	1	10½
				1917.44	353.0	3.95	6.0– 9.0	1	10½
8372	A. Clark 15	2.5 5.2	30 33 33	1919.54	354.7	1.50	5.0–10.0	1	40
				1919.60	356.3	. . . . . .	. .–. . .	1	10½
				1919.63	349.5	1.56	5.0– 9.0	3–2	12
				1922.64	356.2±1.8	1.75	. .–. . .	2	10½
				1926.56	6.5	1.53	5.0–10.0	1	10½
8380	Σ 2281	3.6 7.1	3 58 58	1916.53	77.2±1.5	0.47±0.01	6.3– 7.2	3	40
				1918.68	75.2±1.9	0.58±0.02	6.0– 7.7	5–4	10½
				1919.61	70.2±1.7	0.63±0.04	6.0– 7.8	3	10½
				1919.61	72.8±0.2	0.72±0.10	. .–. . .	2	40
				1919.63	73.7±3.0	0.56±0.02	6.0– 7.4	3	12
				1921.62	67.5±1.6	0.64±0.03	6.0– 7.5	3	10½
				1922.57	67.2±1.0	0.68±0.04	6.0– 7.4	3	10½
				1923.47	63.7±3.2	0.63±0.04	6.0– 7.1	4	10½
				1925.54	63.4±0.4	0.63±0.01	6.0– 7.3	4	10½
				1926.53	61.6±1.7	0.66	6.0– 7.9	3	10½
8390	β 132	4.1 8.2	−19 52 52	1888.50	234.4±1.1	1.00±0.02	7.7– 8.1	2	10
				1892.52	227.8±1.4	0.89±0.06	7.2– 7.3	2	10
				1896.51	224.5±1.6	0.94±0.09	7.0– 7.2	3	10½
				1911.20	217.2±0.5	0.93±0.03	7.0– 7.3	3	10½
				1916.60	214.8±0.2	0.90±0.02	7.0– 7.0	2	12
				1916.64	213.0±2.0	0.83±0.15	7.0– 7.4	3	10½
				1919.59	214.7±2.7	0.97±0.04	6.7– 7.0	3	12
				1919.62	213.0±1.5	1.11±0.00	. .–. . .	2	10½
				1921.70	211.6±1.5	1.06	6.0– 6.5	2–1	10½
		18ʰ 8.2ᵐ	−19°52′	1925.61	210.6±0.2	0.92±0.21	7.0– 7.3	2	10½

BGC	Double Star	R. A. 1880 1950	Dec. 1880 1950	Epoch	Position Angle	Distance	Magnitudes	Nights Observed	Apt.
8393	β 638 AB	18ʰ 4.3ᵐ 7.9	2°34′ 34	1893.51	152.1	21.86	8.4– 9.2	3	16
8393	β 638 BC			1893.50	5.5	1.48	. .–12.2	4	16
8400	Ho 429	5.0 9.0	−15 42 41	1892.56 1911.62	26.4±0.8 24.0±0.9	3.58±0.04 4.08±0.16	8.1–11.2 8.0–11.3	2 3–2	10 10½
8403	Lv 7	5.6 9.6	−15 23 22	1892.53 1911.63 1916.52 1923.39	278.0±2.0 278.2±1.0 279.5±1.3 277.2	3.80±0.14 3.64 3.86±0.00 3.99	8.1–11.7 7.8–11.8 8.5–11.5 8.0–11.5	3–2 2–1 2 1	10 10½ 40 10½
8413	β 292 AB	6.6 10.8	−21 5 4	1892.56	258.6±0.6	17.12±0.06	4.5–11.2	2	10
8413	β 292 AD			1892.56	312.8	48.63	. .– 9.5	1	10
8413	β 292 AE			1892.56	116.0	49.96	. .– 9.7	1	10
8414	β 131 AB	6.7 10.7	−15 38 37	1888.45 1892.53 1911.62 1916.46 1916.52 1917.48 1919.62	278.0 280.2±0.2 279.0±0.4 276.6±1.8 280.3±0.6 277.4±0.4 276.6±2.3	2.61 2.76±0.01 2.62±0.04 3.07±0.18 2.92±0.02 2.91±0.01 2.74±0.18	7.8– 9.8 8.0– 9.5 8.0–10.8 7.2– 9.8 8.0–10.0 7.2– 9.0 8.0–10.2	1 2 2 2 2 2 2	10 10 10½ 10½ 40 10½ 10½
8414	β 131 AC			1916.52 1917.50 1919.63	285.3±0.2 282.1±1.7 281.4	7.90±0.02 8.00±0.04 8.78	. .–11.5 . .–12.4 . .–12.0	2 3–2 1	40 10½ 10½
8414	β 131 AD			1919.62	36.0±1.3	38.91±0.84	. .–11.8	2	10½
8433	Σ 2294	8.4 12.0	0 9 10	1916.53 1921.70 1922.48 1923.52 1925.57 1926.55	99.5±0.6 92.4±0.8 98.4±0.5 97.0 96.7±1.6 96.6±0.4	0.52±0.03 0.56±0.02 0.62±0.04 0.64 0.67±0.00 0.64±0.04	8.0– 8.2 . .–. . . 7.6– 7.8 7.5– 7.7 7.5– 7.6 7.5– 7.7	3 2 2 1 4 2	40 10½ 10½ 10½ 10½ 10½
8448	β 285 AB	9.4 13.7	−25 3 2	1893.55	319.1	1.57	7.5– 9.8	1	16
8456	β 246	10.6 14.7	−19 43 42	1916.63 1917.44 1918.65	102.4±3.8 282.0 310.0	0.48±0.03 0.65 0.63	8.4– 8.1 . .–. . . 8.0– 8.3	4–3 1 1	12 10½ 10½
8458	β 463	10.7 18ʰ 14.8ᵐ	−16 54 −16°53′	1896.50	101.3±1.2	2.22±0.22	9.0– 9.8	4–3	10½

BGC	Double Star	R. A. 1880 1950	Dec. 1880 1950	Epoch	Position Angle	Distance	Magnitudes	Nights Observed	Apt.
8467	Sh 264	18ʰ 11.7ᵐ 15.8	−18°40′ 39	1919.62	51.0±1.3	17.54±0.04	6.5– 8.0	2	12
8483	Schj 16	13.4 14.1	− 5 1 0	1888.48 1921.72	197.2±1.4 199.2±1.0	2.35±0.03 2.46±0.02	7.3– 9.0 7.2– 9.8	3 2	10 10½
8485	Σ 2303	13.6 17.4	− 8 2 1	1885.54 1921.77	228.1 227.2±0.2	2.58 2.32±0.05	7.0– 9.8 6.7–10.0	1 2	10 10½
8488	β 48	13.9 18.0	−19 43 42	1886.30 1892.56 1916.61 1919.55	358.8 0.2 356.9±0.5 355.9	2.51 2.17 2.06±0.06 2.09	. .–. . . 8.2–10.2 8.0–10.3 8.0–10.0	1 1 3–2 1	26 10 12 12
8502	Σ 2306 A & BC	15.4 19.4	−15 9 7	1888.50	220.9	11.84	6.0–. . .	1	10
8502	Δ 18 BC			1888.50	68.0	1.03	7.8– 7.8	1	10
8505	β 1252	15.9 19.8	−11 55 53	1919.63	177.2±0.5	1.45	8.2– 9.5	3–1	10½
8512	Σ 2311	16.6 19.9	11 23 25	1896.48 1922.72 1923.80 1926.70	158.8±0.7 153.1 149.0 149.5±0.9	6.00±0.27 4.72 4.98 4.58±0.17	8.8– 9.6 9.0–10.0 8.8–10.5 9.0–10.3	3 1 2–1 4	10½ 10½ 10½ 10½
8520	β 49 AB	17.0 21.1	−19 38 36	1892.55 1911.59 1916.63 1918.61 1919.55	46.0±0.5 45.4±1.0 44.7±0.1 44.8±0.0 45.6±0.2	8.39±0.16 8.50±0.04 8.26±0.18 7.99±0.00 8.03±0.01	8.3–10.7 8.0–10.3 8.0–11.0 7.9–10.8 8.0–10.2	3 3 3 2 2	10 10½ 12 10½ 12
8520	β 49 AC			1916.69 1918.61	302.6 302.7	23.16 21.35	. .–12.5 . .–12.5	1 1	12 10½
8520	β 49 AD			1916.69 1918.61	149.7 149.6±0.4	24.10 24.24±0.04	. .–12.0 . .–12.0	1 2	12 10½
8535	A.Clark 11	18.7 22.3	− 1 39 37	1916.52	180.0±0.9	0.28±0.02	7.0– 7.2	2	40
8537	OΣ 347	19.0 22.4	7 10 12	1896.52	346.5±0.2	3.13±0.10	7.0–11.5	3	10½
8543	β 1203	20.0 23.6	0 43 45	1916.53	88.8±1.4	0.33±0.01	7.7– 7.9	3	40
8545	Ho 84	20.0 18ʰ 22.8ᵐ	27 20 27°22′	1914.66	321.2±0.8	4.18±0.28	10.1–12.8	4	40

BGC	Double Star	R. A. 1880 1950	Dec. 1880 1950	Epoch	Position Angle	Distance	Magnitudes	Nights Observed	Apt.
8548	Σ 2315	18ʰ 20.2ᵐ 23.0	27°20′ 22	1914.64	181.9	0.24	7.0– 8.5	1	40
8549	β 133	20.2 24.6	−26 42 40	1888.52	265.4±0.4	1.78±0.05	5.8– 6.2	2	10
8569	β 1326	21.8 24.6	26 23 25	1919.56	104.7	5.59	6.5–14.0	1	40
8569	β 1326 AC			1919.60	60.6±0.4	62.17±0.24	7.0– 9.0	2	10½
8571	β 134	22.0 23.9	46 49 51	1896.52	137.3±2.8	1.15±0.05	7.7– 9.7	3	10½
				1916.78	130.5	0.93	8.0– 9.5	1	10½
				1917.84	134.3±0.3	1.14	8.0–10.0	2–1	10½
8590	Howe 43	23.2 27.8	−33 4 2	1892.56	201.0	. . . . . .	6.0–11.0	2–1	10
8617	β 247	25.6 29.4	− 9 27 24	1888.55	168.5±0.3	8.01±0.11	6.5–10.8	2	10
				1911.57	166.4±0.0	8.04±0.10	7.4–10.2	2	10½
				1916.67	167.2	7.83	8.0–10.7	1	40
				1916.68	166.8±1.0	7.94±0.10	8.0–10.7	2	12
8619	β 419	25.7 29.5	− 7 55 52	1888.52	58.0±0.5	1.30±0.20	8.0–10.0	2	10
				1911.17	45.6±0.9	1.14±0.06	8.0– 9.4	3	10½
				1915.73	43.4	. . . . . .	8.0– 9.0	1	10½
				1916.74	40.7±1.3	0.96±0.06	8.0– 9.2	3–2	10½
				1919.77	47.4	1.11	8.3– 8.8	1	10½
8655	Barnard 10	29.8 33.7	−12 5 2	1916.64	134.6	0.43	9.0–10.5	1	40
				1919.54	138.8	0.50	9.0– 9.6	1	40
8663	OΣ 358	30.5 33.6	16 53 56	1911.52	188.6±0.7	1.90±0.02	6.6– 6.9	3	10½
				1914.60	188.5±0.6	1.98±0.02	7.0– 7.2	3	40
				1915.69	187.3±0.7	1.81±0.04	7.3– 7.6	3	10½
				1917.83	186.9±0.7	1.71±0.06	7.0– 7.2	4	10½
				1918.69	187.0±0.2	1.98±0.04	7.2– 7.3	2	10½
				1919.60	186.2±0.5	2.07±0.14	7.0– 7.3	3	10½
				1921.62	185.4±0.4	1.96±0.13	7.0– 7.1	3	10½
				1922.58	186.1±0.5	1.88±0.08	7.0– 7.3	3	10½
				1923.52	184.4±0.4	1.83±0.06	7.0– 7.2	2	10½
				1924.53	184.3±0.6	1.84±0.01	7.0– 7.3	2	10½
				1925.53	184.4±0.5	1.98±0.21	7.0– 7.3	4	10½
				1926.50	184.0±0.6	2.00±0.14	7.0– 7.2	4	10½
8670	β 135	31.3 35.3	−14 6 3	1886.45	187.7	. . . . . .	7.5–12.5	1–0	26
				1892.53	187.1	. . . . . .	8.0–12.0	1–0	10
				1893.54	187.6	2.24	7.0–12.5	2	16
				1916.59	186.2±0.4	2.29±0.00	7.0–12.0	2	40
		18ʰ 35.3ᵐ	−14° 3′	1918.65	188.3±1.6	2.56±0.13	7.1–11.3	4–3	10½

BGC	Double Star	R. A. 1880 1950	Dec. 1880 1950	Epoch	Position Angle	Distance	Magnitudes	Nights Observed	Apt.
8677	Σ 2347	18ʰ 31.8ᵐ	— 0°29′	1886.47	258.8±0.4	3.10±0.22	8.4— 9.8	2	26
		35.4	26	1922.50	257.2±0.0	3.18±0.04	7.5— 9.8	2	10½
8679	A 88	32.1	— 3 18	1916.63	243.0±3.0	0.16±0.00	7.0— 7.3	2	40
		36.1	15						
8679	A 88 AB & C			1916.63	114.4	. . . . . .	. .—. . .	1	40
8709	β 967	34.1	—14 36	1888.45	197.8	. . . . . .	8.0–11.2	1–0	10
		38.1	33	1893.54	191.1	2.33	7.8–11.8	3	16
8710	β 50 AB	34.2	39 29	1916.52	6.3±0.2	21.88±0.31	8.3–12.7	3	40
		36.5	32	1918.71	4.9±1.2	21.82±0.11	8.5–12.6	3	10½
8710	β 50 AC			1916.52	329.3±0.1	73.14±0.23	. .— 9.2	3	40
8710	β 50 CD			1916.52	165.6±0.7	6.06±0.02	. .–11.8	3	40
				1917.81	162.3±3.0	6.65±0.58	. .–11.6	2	10½
				1918.70	167.3±0.1	5.82±0.54	9.7–11.6	2	10½
8736	Σ 2367 AB	36.6	30 11	1916.58	71.8±0.9	0.32±0.01	7.0— 7.5	2	40
		39.3	15	1916.63	70.0	. . . . . .	7.0— 7.5	1–0	12
8736	Σ 2367 AB & C			1916.62	193.0±0.2	14.47±0.16	. .— 8.3	3	12
8740	β 136	37.0	5 37	1892.54	7.7±0.3	4.68±0.04	9.0— 9.2	2	10
		40.4	41	1910.55	6.0±0.4	4.73±0.13	9.0— 9.6	2	10½
				1916.46	6.2±0.2	4.72±0.16	8.8— 9.1	2	10½
				1926.56	6.6±0.3	4.72±0.12	8.8— 9.5	3	10½
13427	A 858	37.6	— 0 21	1916.64	322.4	1.17	9.0–15.0	1	40
		41.2	17						
13428	A 859	37.8	— 0 20	1916.64	15.5	0.26	8.5— 9.0	1	40
		41.4	16						
8750	O. Stone 44	37.8	—20 0	1911.59	288.8±0.8	1.69±0.06	8.5— 8.6	3	10½
		41.9	—19 56						
8751	Σ 2369	37.9	2 30	1896.47	92.7±0.6	1.13±0.03	7.7— 8.3	3	10½
		41.4	34	1911.61	90.7±1.6	1.08±0.00	8.0— 8.5	3	10½
				1921.61	84.6±2.4	0.97±0.06	8.0— 8.3	4	10½
				1925.54	88.7±1.3	0.99±0.05	7.5— 8.8	3	10½
				1926.51	83.7	0.93	7.5— 8.8	1	10½
8755	β 645	38.0	19 21	1896.58	303.6±2.1	9.53±1.4	7.3–11.8	3	10½
		41.1	25						
8770	A 253	39.2	31 34	1914.61	136.3±2.1	0.66±0.04	9.0–10.0	2	40
		18ʰ 41.8ᵐ	31°38′						

BGC	Double Star	R. A. 1880 1950	Dec. 1880 1950	Epoch	Position Angle	Distance	Magnitudes	Nights Observed	Apt.
8776	Σ 2375	18ʰ 39.6ᵐ	5°22′	1888.56	113.2±0.7	2.30±0.08	6.8– 7.4	3	10
		43.0	26	1909.74	114.7±0.2	2.25±0.06	7.0– 7.3	2	10½
				1921.67	116.1±0.4	2.26±0.02	6.0– 6.1	2	10½
				1925.60	115.0±0.2	2.22±0.01	6.0– 6.2	2	10½
8788	β 968 AB	40.6	37 29	1914.75	50.1	. . . . . .	. .–15.5	1–0	40
		43.0	33						
8788	β 968 AC			1914.75	272.4	45.46	4.5–15.0	1	40
8792	Σ 2393	41.1	38 11	1896.47	24.4±0.2	13.38±0.02	7.5– 9.5	3–2	10½
		43.5	15						
8798	Σ 2398	41.6	59 25	1920.83	153.8±0.5	16.86±0.10	8.1– 8.6	5	10½
		42.6	29	1921.73	153.3±0.4	17.11±0.18	8.0– 8.4	4	10½
				1922.74	154.4±0.3	16.83±0.07	8.0– 8.5	3	10½
				1923.52	154.9±0.4	16.70±0.14	8.0– 8.5	4	10½
				1924.62	154.7±0.4	16.82±0.16	8.0– 8.4	4	10½
				1925.58	154.6±0.6	16.58±0.05	8.0– 8.4	4	10½
				1926.65	154.1±0.4	16.83±0.08	8.0– 8.4	4	10½
8798	Σ 2398 AC			1920.83	203.5±0.2	58.80±0.30	. .–11.3	4	10½
				1921.72	202.5±0.3	60.32±0.04	. .–11.1	2	10½
				1922.74	200.5±0.2	62.17±0.19	. .–11.3	3	10½
				1923.52	199.0±0.5	63.08±0.12	. .–11.1	3	10½
				1924.65	197.0±0.2	64.20±0.37	. .–11.3	3	10½
				1925.61	195.2±1.3	65.80±0.10	. .–11.0	2	10½
				1926.65	194.0±0.4	67.44±0.25	. .–11.0	4	10½
8804	β 51 AB	41.7	39 34	1916.51	185.2±0.5	73.72±0.34	9.0–10.5	2	40
		44.0	38						
8804	β 51 BC			1916.51	296.4±0.3	6.48±0.22	. .–11.2	2	40
8837	β 969	43.8	— 8 3	1892.56	238.3±0.4	14.79±0.03	7.5–11.6	3–2	10
		47.6	— 7 59						
8849	β 971	44.4	49 18	1893.52	107.0	0.25 est	. .–. . .	2	16
		46.2	23						
8852	β 265	44.6	11 23	1888.58	236.0±2.5	1.31	7.5– 9.8	2–1	10
		47.9	28	1916.46	236.0±0.4	1.36±0.11	7.2–10.0	2	10½
8887	β 421 AB	48.0	43 15	1893.49	288.5	0.90	9.0– 9.2	2	16
		50.2	20						
8909	β 137 AB	49.8	37 14	1915.78	138.0±1.4	1.37±0.05	7.6– 8.3	4	10½
		52.2	19	1919.66	139.3	. . . . . .	8.0– 8.2	1–0	12
		18ʰ 52.2ᵐ	37°19′	1919.71	139.1±0.6	1.32±0.03	8.0– 8.3	4	10½

BGC	Double Star	R. A. 1880 1950	Dec. 1880 1950	Epoch	Position Angle	Distance	Magnitudes	Nights Observed	Apt.
8909	β 137 AB	18h 52.2m	37°19′	1922.77	138.4±0.6	1.49±0.12	8.0– 8.3	3	10½
				1923.81	137.4	1.42	8.2– 8.4	1	10½
				1924.74	139.7±0.5	1.56±0.08	8.0– 8.4	2	10½
				1925.57	140.8±0.4	1.39±0.01	8.0– 8.3	2	10½
				1926.82	141.4	1.33	8.4– 8.7	1	10½
8909	β 137 AC			1915.79	143.6±0.2	20.14±0.13	. .–12.0	2	10½
				1919.66	143.1	19.98	. .–12.0	1	12
				1919.75	144.8±0.3	19.60±0.36	. .–11.7	3	10½
				1925.57	145.0	. . . . . .	. .–. . .	1–0	10½
8911	β 972 AB	50.0 53.6	— 0 43 38	1893.50	5.4	0.94	8.6– 9.4	2	16
8911	β 972 AB & C			1893.50	14.2	73.48	. .– 9.0	2	16
8933	β 648	52.5 55.1	32 45 50	1893.49	240.2	1.40	6.0– 8.8	2	16
				1896.57	233.4±0.2	1.45±0.13	6.0– 8.8	3	10½
				1919.60	52.1±1.9	1.38±0.07	6.0– 9.8	3	40
				1919.63	49.6±3.6	1.35±0.02	6.2– 9.7	3–2	12
				1919.71	46.5±1.1	1.13±0.11	6.0– 8.6	5	10½
				1920.76	43.8±1.6	1.01±0.08	6.0– 8.0	6	10½
				1921.68	37.3±1.8	1.05±0.09	5.8– 8.3	5	10½
				1922.77	37.4±2.4	0.97±0.06	6.0– 7.8	4	10½
				1923.85	35.0±3.0	0.86±0.06	6.0– 8.0	2	10½
				1924.63	32.6±1.7	0.98±0.06	6.0– 8.0	2	10½
8960½	Hn 33	54.7 59.1	—28 49 44	1888.51	64.8	2.69	8.6– 8.9	1	10
8965	Hd 150	55.0 59.5	—30 3 —29 7	1886.77	271.6	0.47	4.0– 4.3	1	26
				1888.71	261.2	0.65	. .–. . .	1	10
				1891.57	253.7	0.57	. .–. . .	1	10
8966	Σ 2438	55.5 56.7	58 4 10	1896.50	25.0±0.5	0.35±0.02	7.5– 7.5	3	10½
				1920.83	22.2±0.6	0.66±0.01	. .–. . .	2	10½
				1921.69	15.3±1.1	0.62±0.02	7.0– 7.8	3	10½
				1925.57	16.0	0.71	7.0– 7.5	1	10½
8986	Σ 2434 AB	18 56.6 19 0.2	— 0 53 47	1922.59	116.8±0.3	23.43±0.06	7.8– 8.4	3	10½
8986	Σ 2434 BC			1886.41	63.6	1.38	8.8– 9.8	1	26
				1896.47	59.4±0.4	1.32±0.06	8.2– 9.2	2	10½
				1922.48	46.0±3.0	1.25	8.0– 9.8	2–1	10½
. . .	h 5084	18 58.3 19 3.0	—37 14 8	1888.71	188.7	1.2	. .–. . .	1	10
9003	Σ 2444	18 58.5 19h 1.3m	25 53 25°59′	1915.73	314.3±0.5	23.48±0.02	8.6– 9.8	2	10½

BGC	Double Star	R. A. 1880 1950	Dec. 1880 1950	Epoch	Position Angle	Distance	Magnitudes	Nights Observed	Apt.
9009	β 52 AB	19ʰ 1.8ᵐ	25°57′	1896.61	299.8±0.4	51.91±0.07	8.2– 9.6	3	10½
				1915.74	299.2±0.2	51.98±0.03	8.2–10.2	3	10½
9009	β 52 BC	18 58.9	25 51	1896.61	171.0±2.4	8.65±0.21	8.2–11.5	3	10½
		19 1.8	57	1915.74	173.8±0.2	8.78±0.29	10.2–12.2	3	10½
9014	β 466	18 59.6	10 39	1915.72	163.6±0.7	1.54±0.12	9.0–10.3	3	10½
		19 2.9	45	1919.77	166.9	1.97	9.0–10.5	1	10½
				1921.72	164.0±0.4	1.73±0.12	8.5– 9.8	2	10½
9022	S 710	0.0	−16 25	1888.57	1.4	6.27	6.5– 8.3	1	10
		4.0	19	1921.80	1.8±0.6	6.25±0.16	6.3– 8.8	3–2	10½
9024	β 359 AB	0.1	23 15	1919.74	83.1±0.1	4.27±0.07	8.8–10.5	2	10½
		3.0	21						
9024	β 359 AC			1919.75	110.4±0.3	39.56±0.01	. .–10.6	2	10½
9032	Ho 95	0.8	27 6	1916.59	215.6±0.7	0.30±0.03	8.0– 8.4	2	40
		3.6	12						
9038	Σ 2454	1.5	30 15	1916.65	255.8±0.1	0.90±0.00	7.8– 9.2	2	40
		4.2	21						
9059	Ho 98	3.4	26 54	1916.60	133.7	0.21	8.0– 8.3	1	40
		6.2	27 0						
9095	β 1204 AB	6.0	2 25	1916.53	186.3±1.2	0.33±0.02	7.3– 7.6	4	40
		9.5	32						
9095	β 1204 AB & C			1916.53	194.3±0.4	13.10±0.13	. .–14.3	3	40
9095	β 1204 AB & D			1916.53	158.6±0.8	21.41±0.40	. .–14.7	3	40
9095	β 1204 AB & E			1916.55	317.2±0.4	26.56±0.26	. .–14.6	2	40
9095	β 1204 AB & F			1916.55	291.9±0.9	27.38±0.14	. .–14.4	2	40
9106	β 138	6.6	−14 39	1916.61	298.4±3.4	1.24±0.18	7.8–10.8	2	40
		10.6	32	1916.76	299.6±1.6	1.17	8.0–10.5	2–1	10½
9114	Se 2 BC	7.1	38 35	1919.69	96.4±2.4	0.60±0.07	8.0– 8.2	4	10½
		9.5	42	1920.82	92.7±0.3	0.52±0.00	8.0– 8.4	2	10½
				1924.61	88.0	0.45	8.0– 8.0	1	10½
				1925.57	87.6	0.54	. .–. . .	1	10½
9114	Se 2 A & BC			1919.71	217.0±0.4	4.23±0.06	8.2– 8.0	3	10½
				1920.76	216.8±0.7	4.25±0.04	8.0– 7.7	3	10½
				1922.80	215.6	4.07	8.0– 8.0	1	10½
				1924.61	214.9	4.24	8.0– 8.0	1	10½
		19ʰ 9.5ᵐ	38°42′	1925.57	215.0	4.18	. .–. . .	1	10½

BGC	Double Star	R. A. 1880 1950	Dec. 1880 1950	Epoch	Position Angle	Distance	Magnitudes	Nights Observed	Apt.
9116	β 139 AB	19ʰ 7.2ᵐ 10.3	16°39′ 46	1915.80	140.5±1.3	0.72±0.03	7.0– 8.4	2–3	10½
9116	β 139 AC			1915.79	285.0±0.2	114.28±0.20	. .– 8.0	2	10½
9116	β 139 CD			1917.96	261.4±0.5	127.23±0.25	. .–. . .	3	10½
9128	Howe 46	8.2 12.2	−16 11 4	1888.72 1921.83 1925.62	162.0±0.1 159.3±0.3 161.7	5.16±0.04 5.60±0.04 5.41	8.6– 9.2 8.3– 8.6 8.3– 8.6	2 2 1	10 10½ 10½
9154	β 140 AB	10.2 14.1	−11 11 4	1896.66 1916.75 1917.52	324.9±0.6 322.6±0.8 321.2	37.30±0.15 38.56±0.12 39.18	7.9–10.9 7.0–10.8 . .–. . .	3 2 1	10½ 10½ 10½
9154	β 140 BC			1896.66 1916.75 1917.52	208.6±0.3 206.1±0.8 207.9	7.66±0.13 7.28±0.24 7.84	. .–11.6 . .–11.8 11.0–11.6	3 2 1	10½ 10½ 10½
9157	OΣ 368	10.6 13.8	15 57 16 4	1888.55 1921.70	218.0±0.2 215.1±0.8	0.94±0.04 1.00±0.00	7.2– 8.0 7.0– 7.8	3 2	10 10½
9161	S 715	10.8 14.8	−16 10 3	1888.54 1921.80 1925.62	14.9 14.8±0.2 14.7	8.15 8.17±0.12 7.96	6.5– 7.5 7.0– 7.3 7.0– 7.5	1 2 1	10 10½ 10½
9170	S 716	11.2 15.2	−16 10 3	1888.54 1921.80 1925.62	195.6 195.7±0.8 196.2	5.03 4.92±0.08 4.93	8.0– 8.1 8.0– 8.4 8.0– 8.2	1 2 1	10 10½ 10½
9177	Σ 2490	11.7 15.4	− 3 41 34	1886.61 1921.80	248.3±0.1 247.1±1.0	3.40±0.16 3.20±0.00	8.5– 9.8 8.4–10.6	2 2	26 10½
9194	β 248 AB	12.6 15.6	22 49 56	1916.51 1916.64 1926.53	125.0±0.6 125.7±1.2 126.4	2.06±0.04 2.02±0.06 1.67	6.2– 9.5 6.0– 9.8 6.0–10.0	2 3 1	40 10½ 10½
9194	β 248 AC			1916.74	119.7±0.7	51.25±0.12	6.0–11.1	2	10½
9219	Howe 47	14.6 18.1	2 43 50	1888.61 1911.51 1922.75 1925.60	335.6±1.4 332.9±1.1 328.0±1.2 326.9	0.43 0.46±0.03 0.45±0.02 0.46	7.2– 7.2 7.0– 7.0 7.0– 7.5 8.0– 8.1	3–1 3 2 1	10 10½ 10½ 10½
9253	β 141 AB	16.8 19.8	22 17 25	1916.51 1916.80 1917.53	80.7±2.0 79.8±2.0 81.6	0.97±0.00 0.75±0.00 0.75	8.0– 9.5 7.8–10.0 7.5–10.0	2 3–2 1	40 10½ 10½
9253	β 141 CD	19ʰ 19.8ᵐ	22°25′	1916.52	178.9±1.8	5.58±0.38	11.1–12.6	4	40

BGC	Double Star	R. A. 1880 1950	Dec. 1880 1950	Epoch	Position Angle	Distance	Magnitudes	Nights Observed	Apt.
9253	β 141	19ʰ 19.8ᵐ	22°25′	1916.51	332.8±0.4	28.26±0.70	. .–11.2	3	40
	AB & C			1916.78	330.7±0.5	26.72±0.04	7.8–11.8	3–2	10½
				1917.53	329.8	27.58	. .–12.5	1	10½
9253	β 141 AB & E			1916.51	90.6±0.0	50.24±0.08	. .– 9.2	2	40
				1916.76	90.6±0.0	50.15±0.04	. .– 9.7	2	10½
9253	β 141 AB & F			1916.52	213.3±0.5	49.90±0.27	. .–12.8	3	40
9253	β 141 AB & G			1916.51	238.7±2.3	20.84±0.10	. .–15.2	2	40
9279	Σ 2513	19.1	2 13	1886.61	318.6	2.18	8.0– 9.2	1	26
		22.6	21	1921.75	320.6±0.7	2.16±0.07	8.0– 8.9	3	10½
9313	Schj 22	21.5	−12 23	1886.54	327.1	1.56	8.0– 8.2	1	26
	β 142	25.4	15	1888.54	329.0±1.1	1.59±0.09	7.8– 7.9	3	10
				1893.49	330.8	1.53	7.4– 7.9	3	16
				1896.50	333.8±0.8	1.56±0.00	7.7– 8.0	2	10½
				1910.02	342.9±0.3	1.41±0.09	7.9– 8.0	3	10½
				1916.47	347.3	1.42	8.1– 8.0	1	10½
				1918.65	349.5±0.8	1.32±0.09	8.2– 8.1	6–5	10½
				1922.73	352.6	1.33	8.0– 8.1	1	10½
				1926.54	355.7±0.6	1.40±0.08	8.0– 8.3	4	10½
9319	Σ 2525	21.7	27 5	1909.75	311.5±1.2	0.86±0.10	8.0– 8.2	4	10½
		24.5	13	1914.61	308.7±1.3	0.86±0.04	7.3– 7.6	3	40
				1914.79	309.5	0.80	8.0– 8.2	1	10½
				1915.68	306.0±1.6	0.78±0.02	8.0– 8.2	4	10½
				1918.71	306.3±0.6	0.92±0.05	7.9– 8.1	6	10½
				1919.66	303.8	1.06	. .–. .	1	12
				1919.72	304.2±0.3	0.99±0.06	7.5– 7.7	4	10½
				1920.73	305.0±1.3	0.93±0.03	7.5– 8.2	4	10½
				1921.75	303.6±1.3	1.04±0.05	7.7– 7.8	4	10½
				1922.74	304.6±1.1	1.02±0.05	7.5– 7.9	3	10½
				1923.69	303.0±0.7	1.13±0.01	7.5– 7.8	3	10½
				1924.62	302.6±0.5	1.17±0.01	7.5– 7.7	4	10½
				1925.59	302.9±1.2	1.16±0.05	7.5– 7.7	4	10½
				1926.57	303.1±0.3	1.16±0.04	7.5– 7.8	6	10½
9362	Σ 2614	24.8	88 8	1896.59	235.5±0.5	1.61±0.02	9.0– 9.7	3	10½
		20.8	16						
9387	β 143	26.6	49 15	1893.51	192.9	2.15	7.8– 8.7	3	16
		28.5	24	1921.82	192.4±1.4	2.20±0.05	8.0– 9.5	2	10½
9404	β 653 AB	28.2	7 8	1896.50	278.2±0.2	26.56±0.11	4.5–13.0	2	10½
		31.6	17						
9404	β 653 AC	19ʰ 31.6ᵐ	7°17′	1896.50	289.0±0.1	26.58±0.10	. .–13.0	2	10½

BGC	Double Star	R. A. 1880 1950	Dec. 1880 1950	Epoch	Position Angle	Distance	Magnitudes	Nights Observed	Apt.
9424	β 53	19ʰ 29.8ᵐ 33.1	11°11′ 20	1916.62 1916.67	248.6 249.4±4.6	1.53 1.42±0.02	8.8–10.0 8.8–10.0	1 2	40 12
9427	β 655 AC	29.9 30.7	63 3 12	1896.57	288.6±0.1	23.75±0.08	8.0– 8.9	2	10½
9427	β 655 AD			1896.57	275.2±0.5	50.62±0.06	. .– 8.2	2	10½
9434	Σ 2541	30.2 34.1	−10 42 33	1886.41	330.8	4.04	8.5– 9.0	1	26
9447	. . . BC	31.1 34.9	−10 15 6	1888.54	282.8	4.19	9.0–10.0	1	10
9456	Σ 2553	31.8 32.8	61 47 56	1896.55 1921.82	98.7±1.1 102.6±0.8	0.90±0.04 0.84±0.02	8.0– 8.8 8.0– 9.4	3 2	10½ 10½
9464	Σ 2545 AB	32.1 35.9	−10 26 17	1888.52 1922.74 1923.77	321.0±0.7 320.4±1.4 321.5±1.1	3.53±0.07 3.80±0.22 3.55±0.03	6.5– 9.2 6.0– 8.3 . .–. . .	3 4 3	10 10½ 10½
9464	Σ 2545 AC			1922.74 1923.77	166.2±0.5 166.6±0.2	27.22±0.04 27.32±0.35	. .–11.0 . .–. . .	2 2	10½ 10½
9466	β 249 AB	32.2 35.8	0 4 13	1916.62 1916.67 1916.67	132.0±1.4 135.2 135.2±1.6	1.16±0.02 1.30 1.04±0.05	7.3– 9.9 7.3– 9.9 7.5– 9.9	2 1 4	12 40 10½
9466	β 249 AC			1916.67	115.7	18.13	. .–14.0	1	40
9466	β 249 AD			1916.74	56.2±0.09	38.54±0.30	7.2–11.2	4	10½
9481	β 144	33.0 35.8	30 5 14	1915.78 1926.68	351.8±0.2 352.8	6.35±0.01 6.27	9.2– 9.5 9.0– 9.1	2 1	10½ 10½
9500	Σ 2556	34.3 37.3	21 59 22 8	1914.66 1915.77 1916.60 1921.78 1924.60 1925.57 1926.62	130.6±1.0 118.4±1.9 126.7±1.2 115.7 115.4 111.6±1.4 112.3±1.4	0.43±0.01 0.43±0.01 0.35±0.04 . . . . . . 0.48 0.44±0.02 0.44±0.02	7.5– 7.9 6.0– 6.5 7.5– 8.0 . .–. . . . .–. . . . .–. . . 7.1– 7.0	3 3 3 1–0 1 2 4	40 10½ 40 10½ 10½ 10½ 10½
9507	Σ 2557 AB	34.8 37.6	29 28 37	1916.73	105.8±0.0	11.06±0.03	7.2– 9.5	2	10½
9507	Σ 2557 AC			1916.73	303.1±0.3	21.74±0.02	. .–11.5	2	10½
9510	A 166	35.1 19ʰ 38.1ᵐ	23 14 23°23′	1919.68	237.8	0.81	9.0– 9.5	1	40

BGC	Double Star	R. A. 1880 1950	Dec. 1880 1950	Epoch	Position Angle	Distance	Magnitudes	Nights Observed	Apt.
9521	Σ 2564	19ʰ 35.8ᵐ	63°33′	1896.52	170.6±0.8	9.88±0.04	8.2– 8.8	2	10½
		36.6	42	1925.60	164.4	9.06	8.3– 9.7	1	10½
9524	β 145 AB	36.5	30 26	1916.57	264.3	1.04	7.5– 9.5	1	40
		39.3	36	1916.83	274.2±2.8	0.82±0.00	7.4– 9.8	2	10½
				1917.79	270.3±3.2	0.86±0.04	7.2– 9.5	5–4	10½
9524	β 145 AB & C			1916.53	28.9±0.4	8.69±0.02	. .–13.5	2	40
				1916.82	28.0	9.04	. .–12.5	1	10½
				1917.53	30.0	8.96	. .–12.5	1	10½
9524	β 145 AB & D			1916.53	156.4±0.2	26.82±0.06	. .–10.6	2	40
				1916.77	157.6±0.1	26.62±0.14	. .–11.0	2	10½
9547	Collins	37.8	−11 19	1911.20	250.8±0.9	1.35±0.06	8.6– 8.9	3	10½
		41.7	9	1922.44	252.0±1.3	1.43±0.14	8.4– 8.8	4	10½
. . .	56 Sag	39.3	−20 3	1916.84	362.6±1.7	0.42±0.02	6.0– 6.0	3	10½
		43.4	−19 53						
9551	β 827	38.1	−11 29	1911.20	269.7±1.9	0.90±0.08	8.1– 9.0	3	10½
		42.0	19	1921.78	265.6	0.95	8.0– 9.0	1	10½
				1922.62	267.9±0.4	1.01±0.08	8.0– 9.2	2	10½
9574	A. Clark 10	39.2	10 29	1916.52	144.3±2.8	0.28±0.02	7.7– 7.7	2	40
		42.5	39						
9580	β 467	39.4	−21 49	1892.52	134.2	3.13	8.0–11.0	1	10
		43.5	39						
9585	β 468	40.0	3 57	1919.74	184.0±1.0	9.70±0.02	7.2–11.6	2	10½
		43.5	4 7						
9590	β 146	40.1	−20 10	1916.57	319.8	0.86	8.0–10.0	1	12
		44.2	0	1916.73	328.7	0.78	8.0– 9.0	1	10½
				1917.88	310.5	. . . . . .	. .–. . .	1	10½
				1918.64	328.5±3.6	0.98±0.05	8.0– 8.7	3	10½
9594	β 55 AB	40.5	10 16	1916.75	27.8±1.3	3.94±0.20	10.5–10.8	3	10½
		43.8	26						
9594	β 55 AC			1917.84	261.3±0.5	33.67±0.13	9.5– 9.9	3	10½
9602	Σ 2576 AB	41.0	33 20	1915.73	103.2±0.4	2.01±0.08	8.2– 8.3	3	10½
		43.7	30	1919.54	98.2±0.9	1.82±0.16	8.2– 8.4	3	40
				1921.69	95.3±0.6	1.72±0.04	8.1– 8.0	3	10½
				1922.65	94.4±0.2	1.72±0.07	8.0– 8.1	3	10½
				1923.79	93.0±0.5	1.62±0.05	8.0– 8.1	3	10½
				1924.59	92.1±0.6	1.63±0.01	8.0– 8.1	2	10½
		19ʰ 43.7ᵐ	33°30′	1926.64	88.9±0.9	1.40±0.08	8.0– 8.2	2	10½

BGC	Double Star	R. A. 1880 1950	Dec. 1880 1950	Epoch	Position Angle	Distance	Magnitudes	Nights Observed	Apt.
9602	Σ 2576 AC	19h 43.7m	33°30′	1919.55	334.4±0.8	18.25	. .–15.0	2–1	40
9602	Σ 2576 AD			1919.55	255.6±1.0	13.76±0.13	. .–14.0	2	40
9602	Σ 2576 BD			1919.56	256.7	15.91	. .–. . .	1	40
9604	β 828	41.0 44.4	5 52 6 2	1915.74	8.1±0.4	2.48±0.03	8.2–10.2	2	10½
9605	Σ 2579	41.2 43.4	44 50 45 0	1919.83	278.0	1.85	3.0– 7.5	1	10½
				1920.82	277.8±1.1	2.02±0.06	3.0– 7.8	6	10½
				1921.69	276.7±0.6	2.11±0.03	3.0– 7.2	3	10½
				1922.82	276.8±0.2	2.29±0.16	3.0– 7.0	2	10½
				1923.78	275.6±0.0	1.98±0.10	. .–. . .	2	10½
				1924.76	275.0±0.4	2.43	. .–. . .	2–1	10½
				1925.88	273.5	2.29	3.0– 7.0	1	10½
. . .	Jon 150	40.9 44.2	10 6 16	1916.81	193.4±0.2	1.78±0.00	10.2–11.0	2	10½
. . .	Jon 827	42.3 45.4	18 18 28	1919.55	34.1±1.3	3.30±0.02	10.2–11.7	2	40
9623	β 147	42.3 45.0	31 48 58	1893.54	297.6	8.86	8.3– 9.3	3	16
				1910.64	297.7±0.3	8.86±0.08	8.2–10.0	2	10½
9624	Hu 347	42.3 45.4	18 59 19 9	1919.56	341.3±0.6	1.06±0.07	8.8–12.0	2	40
. . .	Jon 1187	42.5 45.6	17 48 58	1919.55	192.0±1.8	1.10±0.08	9.2– 9.6	2	40
9633	β 829	43.0 46.4	5 27 37	1892.52	309.0±1.0	0.80±0.04	8.0– 8.6	2	10
				1896.54	308.6±2.0	0.72±0.04	8.0– 8.7	4	10½
				1922.77	309.2±1.1	0.87±0.17	8.5– 9.0	2	10½
				1926.56	309.7	0.84	8.2– 9.5	1	10½
9634	Σ 2583	43.0 46.3	11 31 41	1888.71	116.0±0.4	1.47	6.0– 6.7	2–1	10
				1909.76	116.1±0.8	1.39±0.08	6.0– 6.7	2	10½
				1913.98	114.7±1.0	1.40±0.05	6.1– 6.6	4	10½
				1926.69	114.8±0.7	1.40±0.08	6.0– 6.5	3	10½
9643	A. Clark 11	43.6 46.7	18 51 19 1	1916.52	173.7±0.6	0.23±0.01	5.5– 6.0	3	40
				1919.56	169.4±0.6	0.30±0.01	. .–. . .	2	40
9650	OΣ 387	44.2 46.8 19h 46.8m	35 0 10 35°10′	1918.72	311.4±0.5	0.67±0.02	6.9– 7.6	4	10½
				1919.73	310.7±2.6	0.62±0.02	7.0– 7.7	3	10½
				1920.85	304.5±3.2	0.64±0.04	7.0– 7.8	4	10½

BGC	Double Star	R. A. 1880 1950	Dec. 1880 1950	Epoch	Position Angle	Distance	Magnitudes	Nights Observed	Apt.
9650	OΣ 387	19ʰ 46.8ᵐ	35°10′	1921.73	302.9±1.1	0.65±0.03	7.0– 7.6	4	10½
				1922.78	301.6±0.7	0.61±0.01	7.0– 7.8	3	10½
				1923.79	302.0±0.9	0.60±0.03	7.0– 8.1	3	10½
				1925.88	295.2	0.55	7.0– 7.5	1	10½
				1926.56	294.6	0.60	7.0– 8.5	1	10½
9654	Hu 349 AB	44.6	16 44	1919.57	235.8	2.21	8.3–14.0	1	40
		47.8	54						
9654	Hu 349 A & CD			1919.57	29.4	53.28	. .–. . .	1	40
9654	Hu 349 CD			1919.57	199.0	3.02	12.5–14.0	1	40
9659	β 361 AB	45.1	22 22	1892.52	350.0±0.8	3.96±0.13	9.0– 9.1	2	10
		48.1	32	1909.74	349.0±0.6	3.74±0.05	9.2– 9.4	2	10½
				1914.64	349.2±0.6	3.75±0.15	9.0– 9.5	2	40
				1919.74	350.2±0.5	3.64±0.02	8.8– 9.4	2	10½
9659	β 361 AC			1919.75	77.8	57.51	. .–11.5	1	10½
9663	β 148 AB	45.4	−10 40	1888.51	326.6	0.78	8.0– 8.5	1	10
		49.2	30	1915.82	303.8±4.4	0.66±0.00	8.0– 8.3	2	10½
				1916.83	308.0±1.5	0.64±0.00	8.1– 8.4	2	10½
				1918.69	302.4±2.2	0.70±0.02	8.0– 8.3	4	10½
9663	β 148 AB & C			1918.72	63.4±0.7	27.70±0.58	8.0–13.0	3	10½
9686	β 979	47.0	22 58	1893.54	335.9	2.25	8.1–11.5	4	16
		50.0	23 8						
9713	Σ 2603	48.6	69 58	1922.74	7.3±0.6	3.31±0.13	4.0– 7.5	2	10½
		48.4	70 8						
9719	Σ 2597	48.9	− 7 3	1888.49	89.3	1.43	7.4– 8.6	1	10
		52.7	− 6 53	1914.69	90.2±1.2	1.26±0.01	7.0– 8.0	2	40
9755	A. Clark 12	52.2	− 2 33	1886.60	325.8±0.0	0.98±0.11	7.6– 8.2	2	26
		55.8	22	1888.54	324.8±2.4	1.16±0.06	6.9– 7.7	3	10
				1909.70	319.4±0.6	0.88±0.04	8.0– 8.4	3	10½
				1914.79	317.3±0.9	0.97±0.03	8.0– 8.8	2	10½
				1921.78	313.7	1.09	8.0– 9.0	1	10½
				1922.50	312.7	1.15	7.8– 9.0	1	10½
				1924.75	313.0±1.1	1.16±0.08	7.5– 8.5	3	10½
9758	β 266	52.2	11 5	1892.57	167.0	15.94	8.0–11.0	1	10
		55.5	16	1909.76	167.2±0.3	16.00±0.02	8.0–11.2	2	10½
				1916.75	165.7±0.4	15.95±0.00	7.0–10.8	2	10½
9759	β 425 AB	52.2	19 58	1919.80	244.7±2.0	1.35	8.7– 9.0	2–1	10½
		19ʰ 55.3ᵐ	20° 9′	1921.68	239.6±0.6	1.40±0.08	8.5– 8.7	2	10½

BGC	Double Star	R. A. 1880 1950	Dec. 1880 1950	Epoch	Position Angle	Distance	Magnitudes	Nights Observed	Apt.
9759	β 425 AB	19ʰ 55.3ᵐ	20° 9′	1919.80	38.8±0.6	21.04±0.22	8.5–11.4	2	10½
	& C			1921.68	38.8±0.8	21.39±0.06	. .–11.8	2	10½
9759	β 425 AB & D			1919.74	85.3	. . . . . .	. .–12.0	1	10½
9769	β 149 AB	52.8	16 10	1893.55	278.6	126.57	6.5– 9.9	3	16
		56.0	21	1915.82	278.4±0.2	128.10±0.11	7.0– 9.1	3–2	10½
				1919.82	278.8±0.0	127.76±0.12	7.0– 9.0	2	10½
9769	β 149 BC			1893.54	199.8	8.32	. .–12.5	4	16
				1915.85	202.3±1.3	8.41±0.17	9.2–11.8	3	10½
				1919.82	202.6±0.4	8.26±0.18	9.0–12.0	2	10½
9792	β 469	54.5	24 24	1919.74	180.4±0.2	14.39±0.11	8.0–10.6	2	10½
		57.5	25	1921.65	180.0±0.8	14.47±0.03	8.0–10.2	2	10½
9833	OΣ 395	56.9	24 36	1914.59	106.1±1.3	0.76±0.03	6.0– 6.5	4	40
		59.9	47	1915.70	106.4±1.2	0.60±0.02	5.6– 6.2	3	10½
				1922.74	105.9±1.3	0.80±0.01	6.0– 6.5	3	10½
				1923.79	107.1±0.6	0.74±0.00	6.0– 6.4	3	10½
				1924.68	107.0±0.1	0.82±0.01	6.0– 6.2	3	10½
				1925.68	106.6±0.2	0.74±0.04	6.0– 6.1	3	10½
				1926.59	107.7±0.5	0.75±0.03	6.0– 6.0	3	10½
9834	Σ 2615	19 57.1	8 4	1896.48	317.2±0.4	10.25±0.06	7.3– 9.6	4–3	10½
		20 0.5	15	1921.72	313.8±0.8	10.08±0.08	7.0– 9.8	2	10½
				1926.74	312.6	10.14	7.0–10.5	1	10½
9864	β 56	19 58.8	− 4 39	1886.70	164.8±0.5	1.55±0.06	8.0– 9.1	2	26
		20 2.5	27	1888.57	166.3	1.70	7.7– 9.0	1	10
				1896.52	165.7±0.9	1.55±0.00	8.0– 8.9	3	10½
				1915.76	168.9±1.2	1.40±0.05	7.6– 9.3	3	10½
				1916.78	167.7	1.61	7.5– 9.5	1	10½
9872	β 426	19 59.2	54 18	1893.50	310.1	5.71	8.1–10.7	3	16
		20 0.9	30						
9873	β 427	19 59.2	54 18	1893.50	335.7	2.93	8.3–10.4	3	16
		20 0.9	30						
. . .	Jon 503	19 58.8	11 18	1918.67	292.0±0.6	4.32±0.14	10.1–10.6	3	10½
		20 2.1	30						
9884	β 57	19 59.9	15 9	1916.53	119.3±0.6	2.66±0.13	6.2–10.7	2	40
		20 3.1	21	1916.81	113.8±0.8	2.61±0.35	6.5–11.2	2	10½
9890	β 832	0.1	−10 59	1888.51	100.1	1.32	8.7– 9.2	1	10
		20ʰ 3.9ᵐ	10°47′						

BGC	Double Star	R. A. 1880 1950	Dec. 1880 1950	Epoch	Position Angle	Distance	Magnitudes	Nights Observed	Apt.
. . .	Espin 203	20ʰ 3.8ᵐ 6.5	35° 7′ 19	1918.75	130.3±1.4	5.80±0.11	8.6–10.2	3	10½
9908	β 428	1.1 4.4	12 36 48	1892.57 1914.59 1914.79	351.4 354.6±1.2 353.6±1.0	0.61 0.73±0.01 0.59	7.5– 8.8 7.0– 8.5 8.0– 9.0	1 2 2–1	10 40 10½
9973	β 150 BC	5.9 8.6	33 17 29	1893.52 1915.94	187.7 185.8±0.1	1.85 1.72±0.06	8.0– 9.8 8.0–10.2	3 2	16 10½
9973	β 150 AB			1915.89	109.2±0.2	41.80±0.10	7.6– 8.2	2	10½
9979	OΣ 400	6.2 8.5	43 35 47	1919.63 1920.88 1921.76 1924.79 1925.88	334.5 337.2 335.5±2.2 332.3±0.8 330.5	0.61 0.69 0.65±0.04 0.61±0.02 0.63	7.0– 7.5 7.0– 7.6 7.0– 7.3 6.0– 6.8 . .–. . .	1 1 5 1 1	10½ 10½ 10½ 10½ 10½
9987	β 430 AB	6.8 9.4	35 28 40	1893.52	21.1	0.98	8.9–10.0	3	16
9987	β 430 AC			1893.55	52.5	17.15	. .– 9.3	1	16
9989	β 982	6.8 9.7	26 1 13	1893.52	50.1	0.81	9.1–10.3	3	16
10005	Σ 2649 AB	7.6 10.4	31 43 55	1896.56 1915.74 1921.78 1923.87	151.8±0.1 151.6±0.0 151.6±0.6 150.8±0.2	24.16±0.02 23.80±0.10 23.26±0.05 23.45±0.06	7.8– 8.9 8.0– 9.2 8.0– 9.0 8.0– 9.4	2 2 3 2	10½ 10½ 10½ 10½
10005	Σ 2649 AC			1921.77 1923.87	284.6±0.1 284.6±0.2	130.58±0.38 130.38±0.02	. .–10.6 . .–10.2	2 2	10½ 10½
10005	Σ 2649 AD			1921.77	37.6	175.33	. .– 9.2	1	10½
10040	β 983	10.2 13.2	25 14 27	1914.65 1922.73 1926.56	160.0±0.6 162.0 162.4	1.18±0.04 1.13 1.29	5.5– 9.8 5.0–10.0 5.0– 8.6	2 1 1	40 10½ 10½
10047	β 59	10.6 14.1	4 45 58	1915.70 1921.78	112.3±0.2 112.2±0.5	8.49±0.18 8.54±0.20	8.9–10.5 8.7–10.5	3 3	10½ 10½
10070	Sh 380	12.5 16.5	−19 30 17	1896.64	177.9±0.2	55.74±0.05	6.0– 8.8	3	10½
10072	β 984	12.5 15.4 20ʰ 15.4ᵐ	26 0 13 26°13′	1888.49 1909.75 1920.87	204.7 210.3±0.7 217.0±1.2	. . . . . . 0.74±0.00 0.80±0.05	. .–. . . 8.0– 8.4 8.0– 8.4	1–0 2 2	10 10½ 10½

BGC	Double Star	R. A. 1880 1950	Dec. 1880 1950	Epoch	Position Angle	Distance	Magnitudes	Nights Observed	Apt.
10072	β 984	20ʰ 15.4ᵐ	26°13′	1922.78	216.1±4.3	0.91±0.01	8.0– 8.7	3	10½
				1923.80	˙216.5	0.72	8.0– 8.6	1	10½
				1924.69	219.6±0.4	0.68±0.02	8.0– 8.5	2	10½
10076	β 441	12.6	28 46	1893.52	65.6	5.93	6.3–11.5	3	16
		15.5	59	1915.74	64.2±1.5	5.74±0.32	7.0–11.4	3	10½
				1919.78	65.2±0.6	5.78±0.08	7.0–11.5	3–2	10½
10106	Barnard 12	14.0	−15 10	1886.77	105.3±0.7	0.98±0.03	7.0–10.2	2	26
		17.9	−14 57						
10113	Schj 25	14.3	− 8 7	1886.60	218.8±1.3	2.82±0.20	8.2– 8.8	2	26
		18.1	− 7 54	1888.70	220.8	2.84	8.2– 9.2	1	10
				1920.88	217.2	2.55	8.2– 9.3	1	10½
				1921.76	219.4	2.62	. .–. . .	1	10½
10148	β 1259 AB	16.4	30 13	1916.63	174.9±1.8	0.44±0.02	8.8– 9.4	3	40
		19.2	26						
10148	β 1259 AB & C			1916.69	9.2	16.00	. .–13.0	1	40
13525	A 1208	17.4	30 52	1916.66	151.3±0.2	0.40±0.04	9.1– 9.4	2	40
		20.2	31 5						
10176	β 664	18.6	5 7	1896.71	287.0±0.6	8.76	7.5–12.2	2–1	10½
		22.1	20						
10207	β 60	20.4	−18 36	1888.61	146.9±0.7	3.53±0.19	5.0– 8.2	5	10
		24.4	23	1915.81	143.7±1.0	3.41±0.03	5.0– 9.5	3	10½
10228	Sh 323 AB	22.0	−18 13	1886.85	174.2	2.72	5.0– 8.0	1	26
		26.0	−17 59	1888.71	174.2	2.78	5.0– 7.4	1	10
				1915.81	171.0±1.3	2.35±0.02	5.0– 7.7	3	10½
				1921.78	168.3	2.24	5.0– 7.8	1	10½
10228	β 61 AC			1915.83	148.5	55.25	5.0–12.5	1	10½
				1916.77	149.3	55.75	5.0–12.0	2–1	10½
10228	β 61 AD Sh 322			1916.77	149.8±0.2	249.49±0.22	. .– 7.0	2	10½
10238	A 293	22.6	41 28	1918.66	120.7	1.60	8.7– 9.0	1	10½
		25.1	42						
10247	β 62 AB	23.1	29 44	1915.80	132.2±0.5	1.07±0.08	8.4– 9.5	3	10½
		26.0	58	1922.73	132.0	. . . . . .	. .–. . .	1–0	10½
10247	β 62 AB & C			1915.79	179.0±0.8	39.16±0.14	. .– 9.7	2	10½
		20ʰ 26.0ᵐ	29°58′	1922.73	179.8	39.09	. .–10.0	1	10½

BGC	Double Star	R. A. 1880 1950	Dec. 1880 1950	Epoch	Position Angle	Distance	Magnitudes	Nights Observed	Apt.
10247	β 62 AB&D	20ʰ 26.0ᵐ	29°58′	1915.79	267.7±0.2	47.84±0.08	. .–11.5	2	10½
				1922.73	268.6	47.42	. .–11.0	1	10½
10264	β 363 AB	24.5	20 12	1919.88	64.1	16.67	7.5–11.5	1	10½
		27.6	26						
10266	β 63 AB	24.6	10 30	1888.58	347.9±0.7	0.88±0.04	5.9– 8.0	3	10
		28.0	44	1909.75	348.2±1.6	0.85±0.04	6.4– 8.2	2	10½
				1914.68	346.4±1.0	0.74±0.06	6.0– 8.2	3	40
				1916.59	344.8±0.8	0.86±0.04	6.4– 8.5	2	12
				1916.61	347.2±0.6	0.72±0.06	6.4– 8.5	2	40
				1916.88	348.8	0.78	6.0– 8.0	1	10½
				1917.33	351.5	0.75	6.0– 8.0	1	10½
10266	β 63 AB & C			1914.69	347.5±0.3	16.82	. .–14.5	2–1	40
				1916.62	347.5±0.9	16.92±0.24	. .–15.3	3	40
. . .	Jon 130	25.7	41 23	1918.66	266.2±0.2	2.22±0.00	9.0– 9.4	2	10½
		28.2	41 37						
10281	Σ 2690 A&BC	25.5	10 51	1914.58	255.9±0.3	16.00±0.10	. .–. . .	5	40
		28.8	11 5	1914.80	256.2±0.6	16.00±0.23	6.5– 7.0	3	10½
				1922.84	255.1±0.3	15.95±0.01	7.0– 7.2	3	10½
10281	Da 1 BC			1888.56	217.2±2.0	0.68±0.12	7.0– 7.0	2	10
				1914.59	212.1±3.1	0.42±0.02	7.5– 7.5	3	40
				1914.80	215.1±5.0	0.43±0.01	7.0– 7.3	3	10½
10281	Da 1 AD			1914.57	105.8±0.5	23.64±0.17	. .–. . .	4	40
10310	β 670	27.3	13 32	1893.52	47.3	0.53	8.5– 8.9	3	16
		30.6	46						
10314	Σ 2696	27.6	5 2	1888.65	298.5	0.85	8.2– 8.4	1	10
		31.1	16						
. . .	Fox 37	28.4	13 1	1918.74	228.6	1.84	9.4– 9.5	1	10½
		31.7	15						
. . .	Espin 247	31.8	36 26	1918.66	143.6	5.47	9.8–11.3	1	10½
		34.5	40						
10363	β 151 AB	31.9	14 11	1893.52	339.2	0.58	4.0– 7.0	2	16
		35.2	25	1914.71	290.8±1.6	0.27±0.02	4.0– 5.0	3	40
				1916.52	317.0±1.6	0.32±0.03	4.0– 5.0	4	40
				1916.89	330.5±3.5	0.35±0.01	4.0– 5.2	3	10½
		20ʰ 35.2ᵐ	14°25′	1917.85	334.5±1.4	0.46±0.04	4.0– 5.4	4	10½

BGC	Double Star	R. A. 1880 1950	Dec. 1880 1950	Epoch	Position Angle	Distance	Magnitudes	Nights Observed	Apt.
10363	β 151 AB	20h 35.2m	14°25′	1918.68	335.6±2.3	0.48±0.04	4.0– 5.3	6	10½
				1919.54	339.7	0.49	. .–. . .	1	12
				1919.56	338.2±1.2	0.49±0.01	. .–. . .	2	40
				1919.74	343.9±1.9	0.45±0.03	5.0– 5.9	5	10½
				1920.79	347.7±1.2	0.53±0.03	5.0– 5.9	4	10½
				1921.75	347.8±0.5	0.60±0.01	5.0– 6.0	7	10½
				1922.67	351.2±0.6	0.60±0.02	5.0– 6.3	4	10½
				1923.78	356.8±1.4	0.62±0.02	5.0– 6.2	2	10½
				1924.73	358.0±0.3	0.59±0.03	5.0– 7.1	6	10½
				1925.68	362.9±0.8	0.61±0.01	5.0– 6.5	3	10½
				1926.63	364.9±2.0	0.56±0.00	5.0– 6.1	3	10½
10363	β 151 AB & C			1914.71	119.1±0.4	23.81±0.11	. .–13.0	3	40
				1916.55	118.6±0.8	23.44±0.27	. .–13.1	3	40
				1917.81	117.0	22.99	. .–12.5	1	10½
				1918.67	119.6±0.2	23.48±0.14	. .–12.7	2	10½
				1919.56	119.6±0.0	23.33	. .–11.5	2–1	40
				1919.73	119.2±0.0	22.98±0.02	. .–12.2	2	10½
10363	β 151 AB & D			1914.71	329.1±0.5	38.47±0.16	. .–10.8	3	40
				1916.88	328.2±0.3	38.48±0.06	. .–10.5	3	10½
				1919.54	328.1	38.43	. .–. . .	1	12
				1919.56	329.0	. . . . . .	. .–. . .	1–0	40
				1919.73	328.6±0.2	38.58±0.22	5.0–10.0	3	10½
10385	β 435	33.2 36.5	14 35 49	1914.64	114.8±1.8	3.09±0.03	8.2–10.8	4–3	40
10427	β 267	35.4 39.1	— 4 49 34	1916.57	62.0±0.4	2.00	9.0– 0.5	2–1	40
10487	β 64 AB	39.3 42.6	12 17 32	1916.55	197.0±2.1	0.30±0.02	8.2– 8.4	4–3	40
				1917.87	203.0±0.2	0.39±0.00	8.1– 8.1	2	10½
10487	β 64 AB & D			1916.56	118.5	62.64	. .–10.5	1	40
				1916.79	118.8±0.1	63.04±0.34	. .–10.3	4–3	10½
10487	β 64 AB & C			1916.79	157.2±0.3	97.14±0.24	7.7– 7.8	4	10½
10488	β 152	39.3 41.0	56 57 57 12	1917.85	97.2±2.0	0.64±0.04	7.0– 8.4	4	10½
10500	β 153	40.2 43.2	—26 51 36	1916.57	270.6	1.50	8.0–10.5	1	12
				1916.78	273.3	. . . . . .	7.5– 9.5	1–0	10½
				1917.84	271.2±1.2	1.38±0.02	7.5– 9.4	2	10½
10519	β 364	41.9 20h 44.9m	24 58 25°13′	1915.74	226.8±1.4	1.14±0.06	8.5– 8.5	3	10½

BGC	Double Star	R. A. 1880 1950	Dec. 1880 1950	Epoch	Position Angle	Distance	Magnitudes	Nights Observed	Apt.
10520	β 65	20ʰ 41.9ᵐ 45.4	5°34′ 49	1888.65	187.5	1.50	5.0– 9.2	1	10
				1909.72	190.4±0.0	1.56	5.0– 9.2	2–1	10½
				1911.60	191.0	1.46	5.0– 9.0	1	10½
				1913.70	188.3	1.35	6.0–10.0	1	10½
				1914.64	191.8±0.8	1.72±0.06	5.2– 9.2	2	40
				1915.76	189.7±0.8	1.40±0.08	6.0– 9.6	2	10½
				1919.68	190.6	2.08	5.0– 8.5	1	40
				1919.77	191.4	1.66	5.0– 8.0	1	10½
10533	OΣ 413 AB	42.7 45.4	36 3 18	1896.57	62.8±1.2	0.62±0.06	5.3– 6.5	5–3	10½
				1906.92	48.9±0.3	0.68±0.06	5.0– 7.0	2	10½
				1910.04	51.5±1.0	0.74±0.01	5.0– 6.5	3–2	10½
				1914.72	57.4±0.3	0.72±0.00	5.0– 6.8	2	40
				1914.80	52.5±1.4	0.58±0.04	5.0– 7.0	3	10½
				1919.74	45.9±2.0	0.59±0.04	5.0– 6.2	4	10½
				1920.82	44.9±1.2	0.68±0.06	5.0– 5.9	4–3	10½
				1921.74	43.5±1.0	0.68±0.05	5.0– 6.0	3	10½
. . .	AG . . .	43.0 46.0	27 9 24	1918.65	19.5±0.2	2.64±0.28	9.0– 9.5	2	10½
10538	β 66	43.0 46.0	27 1 16	1912.66	161.8±0.6	1.01	8.5– 9.0	2–1	10½
				1914.80	163.0±0.4	1.11±0.08	8.4– 9.0	2	10½
10542	β 268	43.2 45.7	41 38 53	1917.86	201.8	0.44	. .–. . .	1	10½
				1918.74	207.1	. . . . . .	. .–. . .	1–0	10½
10557	β 366 AB	44.8 47.0	50 3 18	1893.55	127.5	1.24	8.2– 8.5	2	16
10557	β 366 CD			1893.55	1.2	1.35	10.0–11.0	1	16
10559	Σ 2729	45.1 48.8	— 6 4 — 5 49	1886.75	168.6±0.3	0.47±0.03	6.0– 7.2	2	26
				1914.62	332.2±0.2	0.44±0.02	6.0– 7.0	2	40
				1916.55	333.6±1.0	0.44±0.00	6.0– 7.0	2	40
				1918.44	342.2±0.7	0.50±0.03	6.7– 7.5	3	10½
				1923.81	346.4	0.68	6.0– 7.5	1	10½
				1924.76	343.6±1.6	0.72±0.01	6.0– 6.8	2	10½
				1925.83	350.1	0.75	6.0– 7.5	1	10½
				1926.57	348.2±0.0	0.68±0.00	6.0– 7.2	2	10½
10565	OΣ 415	45.6 48.5	29 58 30 13	1919.90	234.5	3.13	7.0– 9.0	1	10½
10566	β 67	45.6 48.5	30 28 43	1915.93	294.6±1.6	1.38±0.02	7.0– 9.8	2	10½
				1916.73	295.2	1.31	7.0–10.0	1	10½
				1916.69	294.3	1.66	6.7– 9.0	1	40
10569	β 250	45.8 20ʰ 48.2ᵐ	46 13 46°28′	1915.79	4.5±0.8	19.76±0.28	7.3–11.3	3	10½
				1920.82	6.8±0.6	19.26±0.18	7.2–11.5	2	10½

BGC	Double Star	R. A. 1880 1950	Dec. 1880 1950	Epoch	Position Angle	Distance	Magnitudes	Nights Observed	Apt.
10572	h 3003	20ʰ 46.0ᵐ 50.1	—24°14′ —23 59	1886.76	217.4	2.01	6.0– 8.5	1	26
10574	β 154	46.1 50.0	—16 37 21	1886.81 1888.57 1915.90	62.0 66.3 61.0±0.8	2.78 2.71 2.92±0.02	8.3– 8.8 8.5– 9.5 8.2– 9.6	1 1 2	26 10 10½
10581	Ho 144	46.9 50.1	19 41 57	1888.71	165.8	0.48	7.5– 7.5	1	10
10588	β 155 AB	47.4 49.6	50 58 51 14	1916.84 1917.78 1924.79	30.8±1.8 31.1 30.8	0.69 0.63 0.95	6.5– 7.0 6.5– 7.0 6.5– 7.0	2 1 1	10½ 10½ 10½
10588	β 155 AB & C			1916.90 1917.83	24.4 22.0±0.4	17.99 16.00	. .–12.7 . .–12.6	1 2–1	10½ 10½
10588	β 155 AB & D			1917.88	48.2±0.2	199.02±0.18	7.0–11.2	2	10½
10601	Lv 8	49.1 52.9	—11 20 4	1886.71 1892.56 1896.58 1911.60 1916.57 1926.58	298.4±0.3 298.0 291.8±4.2 299.6 301.1 300.9	1.28±0.04 . . . . . . 1.21±0.00 0.96 0.91 1.09	8.4– 9.6 8.5– 9.8 8.8–10.0 8.5– 9.5 8.8–10.5 8.5–10.0	2 1–0 2 1 1 1	26 10 10½ 10½ 12 10½
10607	β 367 AB	49.9 52.9	27 38 54	1916.60	214.4±0.2	0.20±0.00	8.0– 8.5	2	40
10607	β 367 AB & C			1916.59	18.4±0.3	32.08±0.15	. .–12.0	3	40
10607	β 367 AB & D			1916.59	90.3±0.7	28.62±0.20	. .–14.5	3	40
10634	β 764 AB	52.4 56.2	— 9 50 34	1886.79	174.2±0.1	0.57±0.01	9.2– 9.3	2	26
10643	Σ 2737 AB	53.1 56.6	3 50 4 6	1891.61 1914.66	286.0±0.2 307.0	0.88±0.02 0.19	. .–. . . . .–. . .	2 1	10 40
10643	Σ 2737 AB & C			1891.61	74.7±0.3	10.80±0.02	. .–. . .	2	10
10656	β 678 AB	54.3 58.1	— 8 49 33	1886.80	193.9±0.5	. . . . . .	8.3–12.5	1–0	26
10669	β 68	55.6 57.9	49 45 50 1	1915.78	151.6±0.3	1.92±0.08	8.5– 8.8	3–2	10½
10676	Σ 2742	56.3 20ʰ 59.8ᵐ	6 43 6°59′	1888.70 1920.69	224.3 221.0±0.3	2.76 2.63±0.08	7.7– 8.0 6.9– 7.1	1 3	10 10½

BGC	Double Star	R. A. 1880 1950	Dec. 1880 1950	Epoch	Position Angle	Distance	Magnitudes	Nights Observed	Apt.
10685	Σ 2744	20ʰ 57.0ᵐ	1° 4′	1888.53	169.3±0.6	1.50±0.06	6.3– 7.0	4	10
		21 0.6	20	1914.57	157.6±0.4	1.55±0.16	6.0– 7.0	4–3	40
				1922.76	153.0±1.0	1.55±0.03	6.0– 7.0	4–3	10½
				1924.20	152.3±0.5	1.45±0.02	6.0– 7.1	3	10½
				1925.84	150.4±0.2	1.44±0.01	6.0– 6.7	3	10½
				1926.58	151.0	1.60	6.0– 6.5	1	10½
10689	β 69 AB	20 57.2	21 13	1914.62	320.3±1.1	0.89±0.02	7.8– 8.9	3	40
		21 0.3	29	1915.94	314.5±0.9	0.82±0.03	7.7– 8.3	3	10½
				1916.51	319.8±1.8	0.96±0.02	8.0– 8.9	3	40
				1918.77	316.2±1.7	0.80±0.02	. .–. . .	2	10½
10689	β 69 AB & C			1914.66	239.3	77.49	. .–. . .	1	40
				1915.95	239.3±0.0	77.31±0.08	7.8– 7.3	2	10½
				1916.50	239.6±0.2	77.16±0.10	. .– 7.0	2	40
10689	A & C			1918.77	239.4	78.03	7.5– 7.0	1	10½
10689	β 69 CD			1914.60	157.3±0.4	18.82±0.08	7.2–12.8	2	40
				1916.51	158.0±0.8	18.74±0.25	. .–13.2	3	40
				1918.77	156.1±0.4	18.60±0.12	7.0–12.8	2	10½
10690	Σ 2746	20 57.2	38 47	1896.53	294.4±1.8	1.08±0.04	7.8– 8.5	3	10½
		59.9	39 3	1921.75	298.2±1.2	1.06±0.05	8.0– 8.6	3	10½
10692	Lv 9	20 57.4	38 45	1896.58	192.8±1.1	2.35±0.27	9.0–10.6	4	10½
		21 0.1	39 1	1916.69	193.0	2.42	9.3–10.7	1	40
10696	β 156	20 57.6	46 6	1915.95	240.8±0.0	1.02	7.5– 9.8	2–1	10½
		21 0.0	22	1916.78	240.6	1.01	7.0– 9.5	1	10½
10698	Σ 2745	20 57.7	− 6 18	1886.67	191.9±1.1	2.50±0.33	6.5– 8.5	3	26
		21 1.4	2	1921.78	192.7±0.1	2.78±0.02	6.0– 8.1	3–2	10½
10707	β 269	20 58.6	7 17	1916.57	250.1±1.0	1.14±0.12	8.1–10.5	4–2	40
		21 2.0	33	1916.84	252.7	0.89	8.0–10.5	1	10½
				1917.81	251.9	1.01	8.2–10.5	1	10½
10714	β 70 AB	20 58.9	11 33	1915.82	238.5±0.5	79.27±0.44	8.0–10.6	3	10½
		21 2.3	49						
10714	β 70 AC			1915.82	236.1±0.2	74.62±0.12	. .–. . .	2	10½
10714	β 70 BC			1915.82	96.1±1.8	5.78±0.18	11.0–11.1	3–2	10½
10726	A 178	0.5	20 49	1914.72	73.0	0.89	8.0–10.5	1	40
		3.7	21 06						
10727	β 157 AB	0.5	−14 24	1916.75	157.6±0.2	5.57±0.03	7.7–10.8	3	10½
		4.4	7	1921.80	159.4±0.2	5.39±0.23	8.0–10.7	2	10½
10727	β 157 AC	21ʰ 4.4ᵐ	−14° 7′	1916.77	53.6±0.2	7.82±0.07	7.7–11.7	2	10½

BGC	Double Star	R. A. 1880 1950	Dec. 1880 1950	Epoch	Position Angle	Distance	Magnitudes	Nights Observed	Apt.
10731	β 368	21ʰ 1.0ᵐ 4.7	— 8°43′ 26	1886.65 1888.63	90.0±0.6 93.5±0.1	0.66±0.09 0.62±0.04	8.1– 8.6 7.1– 8.0	2 2	26 10
10732	Σ 2758 61 Cyg.	1.2 3.9	38 8 25	1914.96 1920.87	129.5±0.5 131.2±0.2	23.57±0.07 24.19±0.01	5.0– 5.7 5.3– 6.2	2 3	10½ 10½
. . .	AG . . .	3.5 6.5	29 43 58	1918.38	82.5±1.8	5.07±0.04	9.5– 9.8	3	10½
10736	β 473	1.4 5.2	—10 41 24	1893.54 1915.74 1916.84	114.9 114.7±0.3 114.9±1.5	1.88 2.18±0.08 1.83±0.19	8.5–10.0 8.5–10.2 8.3– 9.9	2 2 2	16 10½ 10½
10741	Σ 2759	1.5 4.4	31 58 32 15	1896.52	324.4±0.5	16.06±0.15	8.0– 8.8	2	10½
10743	β 158	1.6 4.0	47 19 36	1915.93 1916.77	314.6 314.4	10.96 11.00	7.5–11.0 7.5–11.0	1 1	10½ 10½
10746	Σ 2760 AB	1.9 4.8	33 39 56	1921.88 1925.90	229.0±0.5 230.6±0.6	3.85±0.06 3.74±0.25	7.1– 7.9 7.0– 7.8	3 3	10½ 10½
10746	Σ 2760 AC			1921.88 1925.90	151.9±0.3 151.8±0.2	59.92±0.24 59.73±0.28	. .–10.2 . .– 9.5	3 3	10½ 10½
10749	OΣ 527	2.0 5.5	4 40 57	1886.77 1888.44 1916.54	283.3±1.5 286.7±8.1 267.2±0.2	0.38±0.02 0.45±0.05 0.39±0.01	7.4– 8.6 . .–. . . 6.0– 8.0	2 2 2	26 10 40
10782	γ Equulei AB	4.5 7.9	9 39 56	1916.54 1916.86 1918.78	270.2±0.4 275.4±1.0 274.5	2.53±0.05 2.16±0.34 2.04	4.0–11.2 4.0–10.7 4.0– 9.8	2 2 1	40 10½ 10½
10782	γ Equulei AC			1916.54 1916.86	5.4±0.1 5.0±0.3	46.28±0.14 46.76±0.48	. .–12.0 . .–12.0	2 4	40 10½
10787	β 251	4.9 9.1	—31 5 —30 48	1916.69 1916.85	232.3 232.3±0.8	2.75 2.52±0.10	7.0–10.0 7.2–10.7	1 3–2	12 10½
10801	H I 47	5.7 9.6	—15 29 12	1888.60 1920.91	321.0 315.9±0.3	3.10 3.28±0.04	7.2– 7.2 7.0– 7.1	1 2	10 10½
10808	β 159 AB	6.4 8.8	47 12 29	1916.68 1916.90	314.4±1.0 313.3±3.2	1.36±0.07 1.08±0.06	6.8– 9.5 6.2– 9.5	2 2	40 10½
10808	β 159 AD			1916.68	144.4	15.44	. .–15.5	1	40
10808	β 159 AE	21ʰ 8.8ᵐ	47°29′	1916.68	337.0	. . . . . .	. .–14.5	1–0	40

BGC	Double Star	R. A. 1880 1950	Dec. 1880 1950	Epoch	Position Angle	Distance	Magnitudes	Nights Observed	Apt.
10818	β 270 AB	21ʰ 7.5ᵐ 11.0	6°43′ 7 0	1916.56 1916.86 1919.70	338.4 347.9 354.5	0.33 0.42 0.42	7.0– 9.5 7.5– 8.5 . .–. . .	1 1 1	40 10½ 10½
10818	β 270 AB & C			1916.57 1916.84 1918.77	30.8±0.2 29.0±2.0 29.8±0.3	32.28±0.13 30.40 32.40±0.06	. .–13.5 . .–12.7 . .–13.0	4 2–1 3–2	40 10½ 10½
10818	β 270 AB & D			1916.78 1918.78	172.6±0.0 172.4	185.16±0.23 185.23	7.2– 7.0 7.5– 7.0	2 1	10½ 10½
10829	OΣ 535 AB	8.6 12.0	9 31 48	1888.58 1914.72 1916.56	212.4±0.8 207.8±1.3 199.6±1.4	0.35±0.00 0.32±0.02 0.29±0.02	. .–. . . 5.0– 5.2 4.0– 4.8	4–2 2 4	10 40 40
10829	OΣ 535 AB & C			1889.73 1890.90 1891.61	20.6±0.4 20.0±0.7 20.6±0.4	40.38±0.10 41.02±0.02 40.99±0.12	. .–. . . . .–. . . . .–. . .	36–12 12–3 24–9	10 10 10
10846	A. Clark 13 AB	10.0 12.8	37 32 49	1914.70 1918.74 1919.71 1920.90 1921.74 1922.78 1924.79 1925.86	211.9±1.6 193.2 188.2±2.6 182.6±0.3 178.6±3.5 175.3±1.3 169.2±0.1 164.4±0.9	0.95±0.11 1.08 1.07±0.10 1.16±0.03 1.13±0.03 1.18±0.10 1.07±0.07 0.98±0.00	5.0– 8.0 4.0– 7.0 4.0– 7.0 5.0– 7.8 4.0– 7.2 4.0– 7.7 4.0– 7.0 4.0– 7.0	3 1 3 2 3 3 2 2	40 10½ 10½ 10½ 10½ 10½ 10½ 10½
10846	A. Clark 13 AC			1914.72	227.7	29.45	. .–13.0	1	40
10849	Σ 2781	10.3 14.0	— 8 9 — 7 52	1886.65 1920.89	170.2±0.4 170.8±0.2	3.11±0.07 2.94±0.04	8.0– 8.2 7.5– 7.7	3 2	26 10½
10855	β 161 AB	10.9 14.6	— 4 45 28	1915.85	350.1±0.2	101.61±0.15	8.2–10.0	3	10½
10855	β 161 BC			1915.85	315.8±0.3	7.41±0.27	10.0–11.6	3	10½
10871	β 162	12.2 15.0	35 16 33	1915.83 1922.76	246.2±0.1 245.3±1.0	1.04±0.04 1.18±0.04	8.2– 8.0 8.0– 8.2	2 2	10½ 10½
10880	β 163 AB	12.8 16.2	11 4 21	1916.64		Round		1	40
10880	β 163 AB & C			1917.90	8.5±0.1	83.53±0.33	7.0–11.3	4–3	10½
10881	β 271 AB	12.8 16.9	—26 51 34	1916.14 1918.76	247.2±1.9 245.7	3.05±0.15 3.51	6.9–11.4 7.0– 9.5	4–3 1	10½ 10½
10881	β 271 AD			1917.88	61.8±0.0	247.24±0.41	7.5–12.0	2	10½
10881	β 271 AE	21ʰ 16.9ᵐ	—26°34′	1918.76	36.8	179.42	. .–12.0	1	10½

BGC	Double Star	R. A. 1880 1950	Dec. 1880 1950	Epoch	Position Angle	Distance	Magnitudes	Nights Observed	Apt.
10884	β 252	21^h 13.0^m 17.1	—27°49′ 32	1888.65 1915.92 1918.76	277.8 273.3±0.7 274.3	2.63 2.36±0.08 2.51	8.2– 8.2 8.2– 8.2 8.1– 8.0	1 3 1·	10 10½ 10½
10910	β 838	14.8 18.3	2 37 55	1896.66 1906.89 1914.80 1915.71	98.7±0.6 98.9 107.2 104.8	1.66±0.15 1.40 1 .19 1.45	8.0– 9.9 8.0–10.0 8.2–10.2 8.0–10.3	3 1 1 1	10½ 10½ 10½ 10½
10914	OΣ 435	15.3 18.8	2 23 41	1915.25	213.0±2.8	0.54±0.06	7.9– 8.3	2	10½
10917	h 281	15.5 18.8	16 14 32	1896.63 1920.96 1921.86	335.3±0.7 334.8 335.0	14.08±0.06 13.89 14.05	8.7– 9.2 9:5–10.7 9.0– 9.5	3 1 1	10½ 10½ 10½
10919	β 1262	15.7 19.6	—15 26 8	1886.76 1920.88	114.0 111.2	1.98 1.96	8.0– 9.2 8.0– 9.5	1 1	26 10½
10947	β 272	17.8 21.6	—13 19 1	1915.91	258.3±0.8	4.59±0.20	8.8–11.4	3	10½
10969	β 164 AB	19.2 22.6	8 52 9 10	1886.78 1915.84 1925.83	238.5±0.5 234.3±0.7 232.2	0.60±0.06 0.60±0.01 0.52	8.0– 8.1 8.1– 8.4 8.0– 8.2	2 3 1	26 10½ 10½
10969	β 164 AB & C			1915.80	241.6±1.0	26.91±0.05	8.2– 9.4	3	10½
13574	A 887	19.6 23.0	10 50 11 8	1916.64	120.2	0.18	8.5– 9.2	1	40
10983	S 790 AB	20.9 23.8	36 9 27	1896.61 1921.94	30.2±0.3 28.1	32.87±0.14 33.23	6.0–10.5 6.0–11.0	2 1	10½ 10½
10983	S 790 AC			1896.61 1921.94	98.2±0.3 98.7	53.90±0.22 54.13	. .– 9.1 . .– 9.0	2 1	10½ 10½
10989	Schj 28	21.4 25.2	—13 57 39	1886.69	132.4±0.2	2.90±0.00	8.1– 8.6	2	26
11001	Σ 2799	23.0 26.4 21^h 26.4^m	10 34 52 10°52′	1888.43 1909.72 1921.78 1922.75 1923.83 1924.72 1925.83	304.6±1.2 294.6±0.2 289.3±0.4 288.6±1.0 289.2±0.2 289.3±0.4 287.3±0.3	1.46±0.06 1.46±0.06 1.49±0.02 1.56±0.14 1.58±0.04 1.56±0.04 1.46±0.07	7.0– 6.7 7.0– 7.0 6.5– 6.6 6.5– 6.5 6.6 –6.5 6.5 –6.5 6.5– 6.5	3 2 3 3 2 3 3	10 10½ 10½ 10½ 10½ 10½ 10½

BGC	Double Star	R. A. 1880 1950	Dec. 1880 1950	Epoch	Position Angle	Distance	Magnitudes	Nights Observed	Apt.
11006	β 72	21ʰ 23.7ᵐ	— 5°55′	1886.62	39.2±2.2	1.88±0.09	8.5–11.3	3	26
		27.4	37	1916.63	41.6±0.4	1.95±0.10	8.7–11.0	3	40
				1916.87	39.7±0.3	1.97	8.7–11.0	2–1	10½
11007	β 684	23.9	— 5 57	1886.78	125.1±0.6	1.06±0.14	9.4– 9.8	2	26
		27.6	39	1916.63	124.2±4.0	1.02±0.05	9.1– 9.5	3	40
				1916.84	122.6	0.79	9.5– 9.8	1	10½
11026	β 73 AB	25.2	— 6 6	1916.82	319.6±0.4	34.92±0.19	3.0–10.8	3	10½
		28.9	— 5 48	1917.84	319.7	34.79	3.0–10.0	1	10½
				1918.76	318.7	35.19	. .–. . .	1	10½
11026	β 73 AC			1916.82	185.1±0.3	57.13±0.26	. .–11.5	3	10½
				1917.83	185.5±0.4	56.62±0.25	3.0–11.4	4	10½
11056	β 165	27.9	— 3 59	1888.82	177.2	5.04	8.3–10.8	1	10
		31.6	41	1920.89	178.6±1.2	5.35±0.30	8.0–10.9	2	10½
11060	β 273	28.6	10 55	1915.91	89.0±0.6	6.02±0.25	8.1–11.3	3	10½
		32.0	11 13						
11068	β 74	29.7	20 52	1893.53	319.9	1.32	7.3– 8.9	3	16
		32.9	21 11	1915.75	322.6±1.6	1.14±0.01	6.4– 9.2	4–3	10½
				1916.82	322.3±0.7	1.16±0.16	6.0– 9.0	3–2	10½
				1917.84	319.6±1.4	1.22±0.10	6.0– 8.9	5–4	10½
11076	β 166	30.3	59 48	1916.83	263.6±3.2	1.06±0.06	7.5–10.5	2	10½
		32.3	60 7						
11077	h 3040	30.4	—20 0	1896.64	47.6±0.3	68.50±0.06	6.0– 9.0	3	10½
		34.3	—19 41						
11088	β 167	31.0	29 31	1916.64	88.8±1.1	2.17	6.8–11.0	3–1	40
		34.1	50	1916.86	84.5±2.1	1.71±0.28	7.2–11.3	3	10½
11125	β 1212 AB	33.3	— 0 36	1914.66	293.5	0.51	7.0– 7.5	1	40
		36.9	17	1914.79	298.1±1.9	0.45±0.06	6.3– 6.8	3	10½
				1916.59	294.1±2.2	0.46±0.02	6.5– 6.9	2	12
				1916.66	291.0	0.45	6.5– 6.9	1	40
11125	β 1212 AB & C			1914.66	146.7	41.44	. .–11.0	1	40
				1914.79	146.1±0.4	41.75±0.30	. .–11.2	3	10½
				1916.61	146.8	41.21	. .–. . .	1	12
				1916.66	146.4	41.43	. .–. . .	1	40
. . .	Espin 520	33.8	27 54	1917.80	36.6±0.7	6.95±0.18	9.7–10.3	3	10½
		36.9	28 13						
11178	β 274	36.4	38 56	1919.85	179.2±2.8	3.46±0.14	8.1–10.8	2	10½
		21ʰ 39.3ᵐ	39°14′						

BGC	Double Star	R. A. 1880 1950	Dec. 1880 1950	Epoch	Position Angle	Distance	Magnitudes	Nights Observed	Apt.
11190	Lv 10	21ʰ 37.1ᵐ 40.9	−11°41′ 22	1886.65	270.0±0.3	1.41±0.12	8.5– 9.6	3	26
				1888.72	271.2±1.2	1.26±0.10	8.2– 9.5	3	10
				1891.82	271.0	. . . . . .	8.5–10.5	1–0	10
				1896.68	273.6±1.8	1.20±0.11	8.5– 9.8	3–2	10½
				1914.57	283.8±0.8	1.21±0.00	8.4– 9.5	2	40
				1916.57	282.8	1.26	8.5–10.2	1	40
				1916.69	281.5	1.24	8.5–10.2	1	12
11210	Ho 166	38.5 41.6	27 18 37	1916.60	51.6	0.32	. .–. . .	1	40
11214	Σ 2822 AB	38.8 41.9	28 12 31	1914.61	131.5±0.6	1.89±0.09	4.0– 5.3	4	40
				1916.63	134.0±0.3	1.75±0.13	4.0– 5.6	2	40
				1916.69	134.8	1.58	4.0– 5.6	1	12
				1919.55	136.1	1.51	. .–. . .	1	40
				1919.74	137.0±0.4	1.53±0.08	5.0– 6.4	3	10½
				1920.79	138.0±1.2	1.37±0.12	5.0– 6.2	4	10½
				1921.81	139.4±0.8	1.40±0.05	5.0– 6.9	5	10½
				1922.76	139.7±0.5	1.36±0.03	5.0– 6.4	3	10½
				1923.80	142.3±0.9	1.25±0.05	5.0– 6.6	3	10½
				1924.72	143.8±1.2	1.22±0.03	5.0– 6.2	4	10½
				1925.85	146.6±0.7	1.06±0.07	5.0– 6.6	5	10½
11214	Σ 2822 AC			1914.60	274.4±0.1	45.57±0.19	. .–11.2	3	40
				1916.68	275.1	46.06	. .–11.2	1	40
				1916.69	274.8	45.51	. .–11.2	1	12
11217	β 374	39.0 41.5	50 27 46	1893.54	141.3	1.86	8.3–10.5	3	16
11222	β 989 AB	39.2 42.4	25 6 25	1886.77		Single		1	26
				1893.51	121.0	0.29	4.0– 4.1	3	16
				1914.73	152.5±6.2	0.16±0.01	. .–. . .	2	40
				1916.58	122.4±1.1	0.24±0.02	4.7– 5.1	4	40
				1919.55	91.4	0.25	. .–. . .	1	40
11222	β 989 AB & C			1914.72	297.4±0.3	12.65	. .–. . .	2–1	40
				1916.59	297.7±0.2	12.56±0.20	4.0–10.8	2	40
				1919.55	294.4	12.48	. .–. . .	1	40
				1919.70	298.1	12.62	5.0–10.0	1	10½
11246	Σ 2825	40.8 44.4	0 18 37	1896.53	113.8±0.6	0.98±0.09	7.8– 8.3	3	10½
				1915.76	113.0±1.0	0.90±0.03	8.1– 8.8	2	10½
11249	Howe 58 AB	41.0 44.8	−13 42 23	1886.76	106.0	0.87	7.8–10.3	1	26
11249	Σ 2826 AB & C	21ʰ 44.8ᵐ	−13°23′	1886.73	81.5±0.5	4.06±0.02	7.9– 8.6	2	26

BGC	Double Star	R. A. 1880 1950	Dec. 1880 1950	Epoch	Position Angle	Distance	Magnitudes	Nights Observed	Apt.
. . .	AG . . .	21ʰ42.1ᵐ 45.2	28°20′ 40	1918.11	159.1±1.8	9.67±0.20	9.7–10.5	3	10½
11272	Σ 2828	43.4 46.9	2 50 3 9	1896.60	141.7±0.5	27.21	8.1– 9.0	3	10½
11275	β 1306 AB	44.0 47.2	23 1 20	1919.68	295.2	31.71	. .—. . .	1	40
11275	β 1306 A & CD			1919.68	275.0	33.30	. .—. . .	1	40
11275	β 1306 CD			1919.68	334.4	. . . . . .	. .—. . .	1–0	40
11317	β 168 A & BC	47.1 51.0	—20 35 15	1915.87	72.8±1.6	5.91±0.12	8.0– 9.5	4–3	10½
11329	β 1213	48.4 51.8	13 0 20	1914.74	311.8	0.72	9.0– 9.5	1	40
11346	β 75 AB	49.7 53.1	10 19 39	1888.68	39.1±1.2	1.27±0.06	8.1– 8.5	2	10
				1896.69	37.8±0.3	1.17±0.08	8.0– 8.2	3	10½
				1906.82	44.6±1.1	1.02±0.07	. .—. . .	3	10½
				1915.75	45.6±0.2	0.80±0.00	8.2– 8.5	2	10½
				1916.67	49.2	0.97	8.0– 8.8	1	40
				1916.73	46.1	0.82	8.0– 8.2	1	10½
				1922.77	50.9	0.83	8.0– 8.3	1	10½
				1924.80	51.2±0.1	0.83±0.14	7.8– 8.0	2	10½
				1925.83	50.5±1.9	0.86±0.09	8.0– 8.6	2	10½
11346	β 75 AB & C			1917.91	212.3±0.6	48.53±0.13	7.8–12.3	3	10½
11350	β 693	49.9 53.6	— 7 33 13	1886.78	49.8	1.05	7.0– 9.8	1	26
11357	S 800	50.4 52.4	62 3 23	1896.54	145.7±0.3	62.84±0.08	6.2– 7.2	2	10½
11369	β 169	50.8 54.7	—21 43 23	1916.86	278.1±0.3	1.82±0.02	9.2– 9.6	2	10½
11385	Σ 2847	51.9 55.5	— 4 4 — 3 44	1886.61	302.6±1.2	1.22±0.02	8.1– 8.4	2	26
11389	h 3074	52.1 55.7	— 2 24 4	1886.84	292.1	2.23	. .—. . .	1	26
				1925.83	293.2	2.16	8.5– 8.9	1	10½
11409	β 275	53.6 21ʰ55.7ᵐ	60 43 61° 3′	1916.84	352.9±3.9	0.43±0.03	7.1– 7.3	4	10½

BGC	Double Star	R. A. 1880 1950	Dec. 1880 1950	Epoch	Position Angle	Distance	Magnitudes	Nights Observed	Apt.
11410	β 276	21ʰ 53.9ᵐ	−29° 2′	1888.90	113.1±1.0	1.48±0.05	5.0– 6.7	3–2	10
		57.9	−28 42	1915.88	114.6±0.6	1.85±0.10	5.5– 7.7	2	10½
11434	S 802	55.9	−17 33	1888.90	244.6	3.88	6.0– 6.1	1	10
		59.7	13	1920.94	244.6	3.78	6.0– 6.5	1	10½
				1921.91	243.0	3.94	7.0– 7.2	1	10½
11443	Ho 610 AB	56.6	26 16	1914.70	238.8±1.1	0.71±0.01	9.0– 9.6	2	40
		21 59.8	36						
11443	Ho 610 AB & C			1914.70	329.2±0.2	32.69±0.07	10.8±0.2	2	40
11443	Ho 610 AB & D			1914.70	282.9±0.5	34.86±0.11	11.1±0.5	2	40
11443	Ho 610 AB & E			1914.71	238.9	57.71	. .–10.7	1	40
11490	Σ 2862	22 1.0	− 0 1	1886.80	100.6±0.6	2.44±0.02	7.8– 8.2	2	26
		4.6	+ 0 19	1923.76	99.8±0.2	2.48±0.04	7.5– 7.6	2	10½
				1925.83	98.3	2.31	7.0– 7.4	1	10½
11512	β 170	2.5	−19 4	1886.54	60.8±0.5	1.69±0.11	8.1– 8.2	4	26
		6.3	−18 44	1888.75	61.8±0.5	1.43	8.4– 8.5	2–1	10
				1916.63	56.6±0.2	1.50±0.12	8.0– 8.2	2	12
				1916.76	54.7±0.5	1.39±0.08	8.1– 8.3	3–2	10½
11529	Lv 11	4.1	−11 40	1886.79	165.1±0.6	0.85±0.02	9.2– 9.6	3–2	26
		7.8	20						
11557	β 475	6.2	− 8 36	1886.76	237.0±0.1	1.57±0.07	7.0–11.2	2	26
		9.9	15						
11580	β 171	7.8	−21 38	1916.07	259.7±1.4	11.63±0.28	8.6–11.3	5–3	10½
		11.7	17						
11590	Σ 2878 AB	8.5	7 23	1886.62	128.2±0.9	1.53±0.08	6.8– 7.9	3	10
		12.0	44	1913.71	126.2±0.7	1.24±0.02	7.6– 8.8	2	10½
				1914.80	127.9±0.8	1.30±0.10	7.2– 8.5	2	10½
11590	Σ 2878 AC			1914.80	123.0±0.3	65.68±0.04	. .– 9.5	2	10½
11592	Σ 2877	8.6	16 36	1914.79	6.0±0.1	12.90±0.15	6.0– 9.0	2	10½
		12.0	57						
11593	β 476	8.7	30 48	1893.54	92.5	2.57	9.4–10.0	4	16
		11.8	31 9	1916.94	94.5	2.70	9.0–10.5	1	10½
				1919.83	93.3	2.35	9.3– 9.7	1	10½
11596	Hd 170	8.8	16 38	1914.79	242.2±0.1	7.40±0.04	10.2–10.6	2	10½
		22ʰ 12.2ᵐ	16°59′						

BGC	Double Star	R. A. 1880 1950	Dec. 1880 1950	Epoch	Position Angle	Distance	Magnitudes	Nights Observed	Apt.
11597	β 991	22ʰ 9.0ᵐ 11.7	51°58′ 52 19	1893.55	143.4	0.57	8.0– 8.2	2	16
11614	β 477	10.5 13.7	30 49 31 10	1893.54 1916.94 1919.83	43.5 43.2 43.4	6.48 6.50 6.26	9.0– 9.8 8.8–10.0 9.2–10.0	4 1 1	16 10½ 10½
11646	h 962 AB	14.4 17.9	5 11 32	1914.61	18.9±0.4	6.11±0.14	6.0–12.5	3	40
11646	h 962 AC			1914.61	224.4±0.06	12.20±0.04	. .–12.1	3	40
11690	Σ 2900 AB	17.9 21.2	20 15 36	1891.79	178.6±0.6	1.83±0.04	6.0–10.4	4–3	10
11691	β 172 AB	17.9 21.6	− 5 27 6	1886.73 1888.66 1914.64 1915.15 1917.85 1920.90 1921.83 1922.77	16.9±1.1 14.8±0.5 354.6±0.4 353.9±1.7 351.5±0.5 354.2±0.9 352.2 351.2	0.71±0.00 0.61±0.05 0.59±0.03 0.62±0.03 0.64±0.04 0.65±0.04 0.66 0.67	6.2– 6.4 5.7– 5.8 . .–. . . 6.2– 6.5 6.0– 6.0 6.0– 6.1 6.0– 6.1 6.0– 6.0	2 3 2 3 2 2 1 1	26 10 40 10½ 10½ 10½ 10½ 10½
11691	β 172 AB & C			1917.85	341.5±0.2	54.57±0.14	. .–11.4	2	10½
11691	β 172 AB & D			1917.85	190.8±0.0	115.75±0.11	. .–10.0	2	10½
11691	β 172 AB & E			1917.85	132.9±0.3	132.52±0.20	. .– 9.1	2	10½
11708	S 808	19.2 23.0	−20 51 30	1896.59	150.8±0.7	6.81±0.05	8.0– 9.0	2	10½
11732	β 291 AB	21.6 25.1	3 55 4 16	1886.77 1916.61 1919.57 1922.77	163.6 176.6±4.1 178.2 179.8	0.32 0.29±0.04 0.42	8.0– 8.3 8.0– 8.6 . .–. . . . .–. . .	1 4 1 1–0	26 40 40 10½
11732	β 291 AB & C			1916.63 1919.56	124.4±0.7 124.4±0.2	30.72±0.04 30.90±0.06	. .–14.8 8.0–14.0	3 2	40 40
11738	β 173	22.4 24.6	56 35 56	1916.84	228.2±3.3	2.63±0.22	8.2–11.1	4	10½
11743	Σ 2909	22.6 26.2 22ʰ 26.2ᵐ	− 0 38 17 − 0°17′	1886.79 1888.95 1906.85 1913.71 1914.97	326.8 325.4±0.4 316.3±0.9 310.6±0.8 308.1±0.7	3.38 3.18±0.02 3.20±0.17 2.95±0.04 2.78±0.07	4.3– 4.0 4.0– 4.0 . .–. . . 4.0– 4.2 4.0– 4.0	1 3 8 2 5	26 10 10½ 10½ 10½

BGC	Double Star	R. A. 1880 1950	Dec. 1880 1950	Epoch	Position Angle	Distance	Magnitudes	Nights Observed	Apt.
11743	Σ 2909	22ʰ 26.2ᵐ	— 0° 17′	1916.53	307.1±1.2	2.99±0.10	4.0– 4.1	3	10½
				1919.53	305.9±0.2	2.99±0.09	. .–. . .	2	40
				1919.87	304.9±0.4	2.72±0.04	4.0– 4.3	4	10½
				1920.90	304.0±0.7	2.77±0.02	4.0– 4.1	3	10½
				1921.95	303.8±0.2	2.83±0.17	4.0– 4.3	3	10½
				1923.81	302.3	2.70	4.0– 4.1	1	10½
				1925.88	302.3±0.5	2.69±0.06	4.0– 4.3	4	10½
				1926.82	301.6	2.59	4.0– 4.4	1	10½
11750	β 174	23.0 26.7	—10 17 — 9 56	1915.82	290.9±1.6	9.75±0.18	8.9–11.4	5	10½
. . .	Jon 856	23.5 26.7	28 43 29 4	1919.68	215.8	1.25	9.5–11.0	1	40
11756	β 76 AB	23.4 27.0	— 0 49 ˉ28	1888.85 1915.86 1919.54	339.0 341.4±1.6 341.7±2.1	1.24 1.22±0.08 1.44±0.09	8.0– 9.7 8.0– 9.9 8.0– 9.5	1 4 3	10 10½ 40
11756	β 76 AC			1919.54	340.0±0.2	45.58±0.25	. .–14.2	4–3	40
11759	β 844	23.5 27.0	5 2 23	1888.87	310.1±2.1	. . . . . .	8.8–11.2	3–0	10
11763	Σ 2912	23.9 27.4	3 49 4 10	1886.77 1916.62	168.0 282.2	0.34 0.23	7.5– 8.8 . .–. . .	1 1	26 40
11793	Ho 475 AB	27.1 30.4	25 48 26 9	1914.70	141.3±1.0	0.84±0.8	8.3– 8.4	3	40
11793	Ho 475 AB & C			1914.70	223.9±0.5	8.14±0.04	. .–11.3	3	40
11803	β 77 AB	27.8 31.4	— 2 24 3	1886.81 1888.72 1914.63 1915.93	212.2±0.2 213.8±0.7 216.1±1.8 211.9±0.8	2.64±0.02 2.45±0.11 2.70±0.02 2.50±0.07	8.2– 8.8 8.2– 9.3 9.0– 9.6 8.8– 9.8	2 2 3 3	26 10 40 10½
11803	β 77 AC			1914.63 1915.95	226.0±0.3 224.6±0.2	27.88±0.17 28.11±0.00	. .–11.3 8.8–11.0	3 2	40 10½
13621	Hu 982	29.3 32.7	14 0 22	1916.59	212.7±0.7	0.92±0.04	7.2–10.7	2	40
11888	β 277 AB	34.2 37.3	40 45 41 7	1893.55 1916.69 1916.87	201.9 208.0 200.5±1.4	0.49 0.53 0.63±0.04	8.0– 8.4 8.0– 8.5 8.1– 8.0	1 1 3–4	16 12 10½
11888	β 277 AB & C			1916.92 1917.91	114.6 115.6±0.7	25.88 25.60±0.01	. .–11.5 7.0–12.0	1 3–2	10½ 10½
11888	β 277 AB & D	22ʰ 37.3ᵐ	41° 7′	1917.93	46.5±0.6	75.48±0.12	7.6– 7.8	3	10½

BGC	Double Star	R. A. 1880 / 1950	Dec. 1880 / 1950	Epoch	Position Angle	Distance	Magnitudes	Nights Observed	Apt.
11895	Ho 296	22ʰ 34.9ᵐ / 38.3	13°55′ / 14 17	1914.73 / 1916.58	40.6±2.8 / 29.8±1.8	0.40±0.02 / 0.28±0.00	5.2– 6.0 / 6.0– 6.5	2 / 3	40 / 40
11903	β 709	35.4 / 39.0	— 3 11 / — 2 49	1886.80 / 1888.89 / 1896.81 / 1921.87	6.6±0.6 / 5.7 / 5.4±1.2 / 4.2±1.7	2.01±0.03 / 1.51 / 2.04±0.11 / 1.97±0.07	8.2– 9.0 / 8.3– 9.5 / 8.3– 9.4 / 8.2– 8.8	2 / 1 / 3 / 2	26 / 10 / 10½ / 10½
11914	Σ 2935	36.8 / 40.5	— 8 56 / 34	1886.87 / 1920.94 / 1921.86	310.3±0.6 / 308.8±1.5 / 309.7	2.56±0.12 / 2.63±0.34 / 2.22	6.4– 8.2 / 6.5– 7.7 / . .–. . .	2 / 2 / 1	26 / 10½ / 10½
11917	β 710	37.0 / 40.3	29 5 / 27	1916.63	232.9±3.2	0.44±0.03	8.9– 9.2	4	40
11920	β 176	37.1 / 40.2	38 40 / 39 2	1915.77	43.2±0.6	2.06±0.06	9.1– 9.5	2	10½
11924	β 1144 BC	37.4 / 40.7	29 36 / 58	1916.60	62.7	0.35	9.7–10.2	1	40
11924	β 1144 A & BC			1916.66	338.8±0.0	91.00±0.12	4.0–11.0	2	40
11924	β 1144 DE			1916.64	181.4	5.70	14.0–16.0	1	40
11924	β 1144 BC & D			1916.64	320.4	62.86	. .–. . .	1	40
11930	OΣ 477	38.3 / 41.3	45 24 / 46	1896.74	190.4±3.2	4.45±0.13	7.3–11.0	3	10½
11936	β 450 AB	38.7 / 41.7	38 50 / 39 12	1915.79	279.9±2.4	2.80±0.08	6.5– 9.0	3–2	10½
11936	β 450 AC			1893.54 / 1915.75	231.8 / 233.8	10.86 / 11.56	7.0–12.5 / . .–12.5	1 / 1	16 / 10½
11957	h 301	40.7 / 44.2	11 33 / 55	1896.70	110.5±0.8	12.30±0.06	4.0–12.3	3–2	10½
11968	Σ 2944 AB	41.7 / 45.3	— 4 51 / 29	1886.77 / 1888.90 / 1906.93 / 1920.95 / 1921.86 / 1926.00	254.6±0.1 / 256.6 / 259.0±0.6 / 260.5 / 262.5±1.0 / 264.4	3.41±0.11 / 3.26 / 3.20±0.02 / 3.00 / 3.04±0.09 / 2.88	7.4– 7.6 / 7.6– 7.8 / 7.0– 7.2 / 7.0– 7.2 / 7.0– 7.3 / 7.0– 7.2	2 / 1 / 2 / 1 / 2 / 1	26 / 1 0 / 10½ / 10½ / 10½ / 10½
11968	Σ 2944 AC			1906.93 / 1921.86 / 1926.00	127.4±0.1 / 121.4±0.2 / 118.6	47.78±0.05 / 47.45±0.22 / 47.17	. .– 8.5 / . .– 8.0 / . .– 8.5	2 / 2 / 1	10½ / 10½ / 10½
11981	Hn 53	42.9 / 22ʰ 46.5ᵐ	— 7 8 / — 6°46′	1886.46	2.1±0.1	1.72±0.06	8.8–10.2	3–2	26

BGC	Double Star	R. A. 1880 1950	Dec. 1880 1950	Epoch	Position Angle	Distance	Magnitudes	Nights Observed	Apt.
12008	Ho 482 AB	22ʰ 45.7ᵐ	25°45′	1914.73	205.0±2.2	0.22±0.01	7.2– 7.8	2	40
		49.0	26 7	1919.58	192.5	0.28	. .–. . .	1	40
12008	Ho 482 AB			1914.73	198.6±0.4	. . . . . .	. .–. . .	1–0	40
	& C			1914.79	198.3±0.2	51.06±0.04	6.5– 9.4	2	10½
12012	β 177	45.9	−22 21	1888.84	277.0±2.2	2.62±0.14	8.2– 8.3	2	10
		49.7	−21 59						
. . .	Vander-donck 2	48.4	7 46	1918.74	127.8±4.6	3.32±0.12	9.7– 9.7	2	10½
		51.9	8 8						
12036	β 382	48.3	44 7	1893.53	221.3	0.91	6.2– 8.0	3	16
		51.4	29						
12046	β 178	49.0	− 5 38	1886.78	320.5±2.8	0.77±0.06	6.0– 8.5	4–3	26
		52.6	16	1888.87	322.9±1.5	0.70±0.05	6.0– 8.0	3–2	10
				1914.60	327.7	0.63	8.0– 9.5	1	40
				1914.79	335.2±0.6	0.46±0.02	6.0– 7.5	2	10½
				1916.61	325.7±2.0	0.40±0.04	6.0– 8.0	2	40
				1916.69	326.3	0.44	6.0– 8.0	1	12
				1917.86	340.3±0.3	0.48±0.06	6.0– 7.5	2	10½
12051	β 1010	49.3	− 6 13	1886.55	. . . . . .	1.18	9.0– 9.2	0–1	26
		52.9	− 5 51						
12058	β 383 AB	50.0	8 49	1914.63	117.8±0.8	2.48±0.13	7.5–13.5	3	40
		53.5	9 11						
12058	β 383 AC			1914.63	241.5±0.5	15.89±0.32	. .–11.5	3	40
12069	Σ 2959 AB	50.9	− 3 53	1914.58	107.2±0.3	12.56±0.09	6.6–10.0	4	40
		54.5	31						
12069	Σ 2959 AC			1914.54	102.2	22.47	. .–13.0	1	40
12069	Σ 2959 BC			1914.60	95.4±0.4	10.88±0.01	10.0–13.2	3–2	40
12090	OΣ 536	52.5	8 43	1896.55	170.5±8.2	0.36±0.05	. .–. . .	3	10½
		56.0	9 5	1925.85	28.0±9.5	0.38±0.02	. .–. . .	2	10½
12094	OΣ 483	53.2	11 5	1914.58	231.9±1.3	0.98±0.05	6.5– 8.0	3	40
		56.7	27	1916.58	234.0±1.2	0.99±0.09	6.2– 8.4	5	40
				1919.58	234.4±2.2	1.00±0.05	6.0– 7.2	4	40
				1919.79	237.4±0.8	0.80±0.00	6.0– 7.5	2	10½
				1920.83	241.0±2.0	0.87±0.07	6.0– 7.6	5–4	10½
				1921.82	240.3±1.5	0.91±0.05	6.3– 7.8	3	10½
				1922.78	238.9±0.6	0.98±0.09	6.0– 7.6	3	10½
		22ʰ 56.7ᵐ	11°27′	1925.85	243.6±0.6	0.88±0.03	6.0– 7.6	2	10½

BGC	Double Star	R. A. 1880 / 1950	Dec. 1880 / 1950	Epoch	Position Angle	Distance	Magnitudes	Nights Observed	Apt.
12108	β 179	22ʰ 54.4ᵐ / 58.2	−22°54′ / 32	1915.78	114.5±0.3	13.00±0.03	8.6– 9.1	3–2	10½
12118	β 384	56.2 / 59.9	−19 10 / −18 48	1886.81 / 1888.85	70.7 / 74.5	1.17 / 1.01	6.8– 8.8 / 9.4– 7.0	1 / 1	26 / 10
12119	β 481	22 56.4 / 23 0.1	−11 53 / 31	1886.74	55.1	. . . . . .	9.5–10.5	1–0	26
12134	h 1842	22 57.9 / 23 1.2	27 26 / 49	1896.68	209.2±0.4	102.87±0.11	3.7–10.6	3	10½
12146	Howe 62	22 59.3 / 23 2.8	− 4 54 / 31	1888.89 / 1922.76	211.7 / 212.3±0.9	4.03 / 4.80±0.12	8.0–11.0 / 8.1–11.0	1 / 2	10 / 10½
12183	Σ 2980	3.0 / 6.6	− 7 58 / 35	1886.86 / 1906.81 / 1922.76	108.4±0.2 / 108.4±0.6 / 107.0±1.8	4.50±0.06 / 4.58±0.16 / 4.67±0.03	7.0– 9.8 / . .–. . . / 7.5–10.9	2 / 2 / 2	26 / 10½ / 10½
12185	Σ 2981	3.2 / 6.8	− 9 29 / 6	1886.74 / 1906.79 / 1915.44	113.4±0.1 / 110.4±0.5 / 111.0±1.6	3.76±0.08 / 3.80±0.12 / 3.79±0.10	8.2– 8.5 / 9.0– 9.0 / 9.1– 9.4	3 / 2 / 3	26 / 10½ / 10½
12196	OΣ 489	4.1 / 6.4	74 44 / 75 7	1896.79	35.8±2.9	0.90±0.12	5.8– 7.8	2	10½
12205	β 852 BD	4.8 / 8.2	25 52 / 26 15	1914.72	209.8±0.7	19.31±0.35	. .–14.7	3	40
12205	β 852 BC			1914.73	359.9±1.1	1.48±0.10	10.7–11.4	4	40
12205	β 852 AB			1914.74	282.6±0.1	58.38±0.03	7.0–10.7	2	40
12213	Σ 2988	5.7 / 9.4	−12 35 / 12	1886.72 / 1921.95	279.7±0.4 / 278.3±1.4	3.66±0.10 / 3.52±0.02	7.9– 8.0 / 7.2– 7.3	2 / 3	26 / 10½
12231	β 181 AB	7.5 / 11.2	−14 3 / −13 40	1886.82 / 1888.77 / 1916.37	308.2±1.6 / 305.2±0.6 / 308.0±1.2	1.51±0.06 / 1.30 / 1.24±0.10	6.9– 9.5 / 7.2– 9.8 / 7.0–10.2	4–2 / 3–1 / 2	26 / 10 / 10½
12231	β 181 AC			1916.24	240.8±0.3	19.45±0.09	7.2–11.7	3	10½
12237	β 714	7.9 / 11.5	− 3 17 / − 2 54	1886.73	146.2±0.2	0.42±0.02	7.1–10.4	2	26
12242	β 715	8.4 / 12.1	−11 20 / −10 17	1886.65		Single		1	26
12257	β 1220 BC	9.6 / 13.2	− 9 44 / 21	1916.64	279.7	0.62	9.2– 9.4	1	40
12257	β 1220 BC & E	23ʰ 13.2ᵐ	− 9°21′	1916.64	348.0	18.66	. .–. . .	1	40

BGC	Double Star	R. A. 1880 1950	Dec. 1880 1950	Epoch	Position Angle	Distance	Magnitudes	Nights Observed	Apt.
12274	β 182 AB	23ʰ 10.9ᵐ	−14°28′	1886.76	45.9±0.9	0.74±0.04	8.0– 8.2	3–2	26
		14.6	5	1888.77	47.0±2.7	0.72±0.06	8.2– 8.2	2	10
				1917.86		Single		2	10½
12274	β 182 AB & D			1916.89	165.1±0.4	141.40±0.36	8.1–11.1	4	10½
				1917.92	164.8	140.41	7.0–11.0	1	10½
12276	β 79	11.4	− 2 10	1886.61	101.2±1.0	0.96±0.06	7.5– 9.2	2	26
		15.0	− 1 47	1893.55	87.7	0.86	8.0– 9.0	1	16
				1915.84	66.6±0.8	0.92±0.04	8.0– 9.9	2–3	10½
				1916.63	71.8±0.1	1.12	8.0– 9.8	2–1	40
				1916.69	67.3	1.10	8.0– 9.8	1	12
				1917.86	66.3	1.07	8.0–10.5	1	10½
				1919.66	65.3	1.18	8.0– 9.7	1	40
				1922.77	58.2±1.0	1.05±0.01	8.0– 9.8	2	10½
12290	β 80 AB	12.8	4 45	1886.94	315.6	0.76	7.7– 8.7	1	26
		16.4	5 8	1888.76	319.6±1.6	0.84±0.05	8.0– 9.1	3	10
				1893.55	328.0	0.81	8.4– 9.0	1	16
				1914.74	240.0	0.63	8.3– 9.8	1	40
				1915.90	240.8±0.0	0.56±0.02	8.2– 9.1	2	10½
				1916.81	244.0±2.3	0.63±0.03	8.0– 9.2	3	10½
				1917.84	248.8±2.0	0.75±0.02	8.0– 9.0	4–3	10½
				1919.67	248.3±2.5	0.80±0.01	8.0– 9.0	2	40
				1919.70	251.2	0.83	8.0– 9.3	1	10½
				1920.85	252.6±0.9	0.82±0.06	8.2– 9.2	3	10½
				1922.78	254.2±2.1	0.87±0.04	8.0– 9.3	3	10½
				1923.81	253.2	0.91	8.0– 9.5	1	10½
				1924.81	259.4	1.01	8.0– 9.0	1	10½
				1925.88	258.3	0.91	8.4– 9.0	1	10½
12290	β 80 AB & C			1916.83	359.5±0.1	104.80±0.16	8.0– 9.9	3	10½
				1922.80	358.3	105.92	8.0–10.0	1	10½
12290	β 80 AB & D			1916.83	330.4±0.1	192.35±0.30	8.0– 8.9	3	10½
				1922.80	329.6	194.39	8.0– 9.0	1	10½
12290	β 80 AB & E			1916.84	294.6±0.2	205.50±0.23	8.0– 9.4	2	10½
12290	β 80 CE			1916.83	134.9±1.6	27.12±0.36	10.0–11.8	2	10½
12296	h 5394	13.2	− 5 47	1896.77	21.3±0.1	10.57±0.06	5.2–10.5	3	10½
		16.8	24						
. . .	Fox 47	11.9	9 48	1917.87	280.5±0.6	4.74±0.12	8.8–10.4	2	10½
		23ʰ 15.4ᵐ	10° 1′						

BGC	Double Star	R. A. 1880 / 1950	Dec. 1880 / 1950	Epoch	Position Angle	Distance	Magnitudes	Nights Observed	Apt.
12311	Ho 488 A & BC	23ʰ 14.7ᵐ / 18.3	1°48′ / 2 11	1914.62	201.1±1.3	4.26±0.12	. .–. . .	4	40
12311	Ho 488 BC			1914.62	215.7±1.8	0.85±0.08	. .–10.7	3	40
12332	Σ 3007 AB	16.8 / 20.3	19 54 / 20 17	1914.92 / 1921.93	86.0±0.4 / 85.0±0.9	6.16±0.12 / 5.83±0.14	7.0–10.5 / 7.0–10.3	2 / 3	10½ / 10½
12332	Σ 3007 AC			1914.92 / 1921.93	318.6±0.2 / 317.1±0.3	79.78±0.02 / 81.57±0.24	. .–10.4 / . .–10.3	2 / 3	10½ / 10½
. . .	Jon 296	17.1 / 20.7	6 5 / 28	1917.91	137.2±0.2	4.49±0.22	9.6–10.2	2	10½
12340	Σ 3008	17.5 / 21.1	− 9 7 / − 8 44	1888.79 / 1921.83	251.3±1.2 / 218.8±0.4	4.28±0.01 / 3.40±0.08	7.2– 8.2 / 7.0– 8.0	2 / 2	10 / 10½
12345	β 854	18.2 / 21.8	5 23 / 46	1888.82	88.8	1.86	8.5– 8.5	1	10
12363	Ho 489 AB	20.1 / 23.5	27 3 / 26	1914.71	242.5±0.6	0.55±0.03	8.0– 8.3	4–3	40
12404	β 1266 AB	24.5 / 27.9	30 10 / 33	1893.55 / 1916.60	65.9 / Round with power 700	0.23	Equal	1 / 1	16 / 40
12404	β 1266 AB & C			1916.63	203.3±0.1	18.83±0.13	7.0– 9.0	2	40
12404	β 1266 AB & D			1916.66	7.7	36.82	. .–14.5	1	40
12408	OΣ 497	24.8 / 28.3	8 49 / 9 12	1888.56 / 1922.76	213.1 / 213.4±1.1	1.24 / 1.22±0.02	8.0– 8.5 / 8.0– 9.2	1 / 2	10 / 10½
12432	β 720	28.0 / 31.5	30 40 / 31 3	1893.55 / 1896.67 / 1914.65 / 1916.64 / 1919.53 / 1921.86 / 1922.77 / 1926.78	151.3 / 157.0±4.6 / 184.9±1.7 / 188.4±0.8 / 194.0±0.2 / 190.4±0.6 / 192.5 / 200.6	0.30 / 0.39±0.01 / 0.42±0.01 / 0.36±0.02 / 0.46±0.02 / 0.46±0.03 / 0.44 / 0.43	6.5– 6.0 / 6.0– 6.5 / 6.0– 6.0 / 6.0– 6.1 / . .–. . . / 6.0– 6.3 / . .–. . . / 6.0– 6.5	1 / 3 / 3 / 3 / 2 / 3 / 1 / 1	16 / 10½ / 40 / 40 / 40 / 10½ / 10½ / 10½
12443	β 81	29.0 / 32.6	−12 14 / −11 51	1886.11 / 1915.86	14.6±0.4 / 8.7±1.3	1.94±0.06 / 1.51±0.17	8.2–11.0 / 8.4–10.5	3 / 3	26 / 10½
12450	β 721	30.1 / 33.7	− 7 47 / 24	1886.81	133.9±0.4	0.30±0.04	8.0– 8.3	2	26
12457	h 3206	30.8 / 23ʰ 34.5ᵐ	−22 20 / −21°57′	1886.88	348.0±1.3	3.47±0.32	8.2– 8.4	4–3	26

BGC	Double Star	R. A. 1880 1950	Dec. 1880 1950	Epoch	Position Angle	Distance	Magnitudes	Nights Observed	Apt.
12498	β 723	23ʰ 34.5ᵐ 38.1	− 0°15′ + 0 8	1888.93	168.9	. . . . . .	7.0–12.0	1–0	10
12500	Σ 3030	34.6 38.2	− 1 3 − 0 40	1886.80 1920.88 1921.83	222.6±0.3 223.4 222.2	2.42±0.02 2.38 2.41	8.2– 8.6 8.0– 8.2 . .–. . .	2 1 1	26 10½ 10½
12501	Ho 303	34.8 38.3	19 42 20 5	1914.73	192.2±0.4	0.65	8.0–12.0	3–1	40
12523	β 279	36.5 40.1	−15 12 −14 49	1888.93 1916.69 1916.70 1916.82	86.2±0.8 85.0 87.0 85.1±0.4	5.08 5.86 5.51 5.37±0.32	5.0–10.8 5.0– 9.8 5.0– 9.8 5.0–10.4	2–1 1 1 5	10 12 40 10½
12543	Sh 356	39.8 43.4	−19 21 −18 58	1888.88	139.3±0.0	5.90±0.01	6.0– 7.1	2	10
12544	Σ 3036	39.9 43.5	− 0 24 1	1886.81	228.0±0.4	2.48±0.11	7.9–10.4	3–2	26
12549	β 726	40.4 44.0	−13 25 2	1886.86	326.7±2.8	0.62±0.05	8.1–10.2	3	26
12601	Σ 3042	45.9 49.4	37 14 37	1914.87 1921.87	85.7±0.4 87.8±0.3	4.88±0.03 4.92±0.05	7.0– 7.1 7.0– 7.2	2 2	10½ 10½
12606	Ho 204	46.2 49.7	27 55 28 18	1896.78 1921.44	356.2±0.5 355.4±0.4	5.87±0.05 5.83±0.03	8.0– 9.9 8.0–10.5	3 2	10½ 10½
12609	β 859	46.6 50.1	22 18 41	1914.57	211.8±0.4	0.63±0.05	8.5– 8.6	2	40
12624	Sh 358	48.2 51.7	31 14 37	1896.81 1905.97 1914.70	329.2±0.2 329.1 120.5±1.4	37.47±0.12 37.47 0.86±0.04	7.8– 8.9 . .–. . . 6.0–11.0	2 1 4–2	10½ 10½ 40
12639	Σ 3046	50.2 53.8	−10 10 − 9 47	1886.73 1888.50 1907.90 1914.84	247.0±0.5 248.2±0.5 249.6±1.0 351.6±0.8	3.02±0.12 2.96±0.13 2.95±0.18 3.09±0.03	8.0– 8.6 7.9– 8.4 8.4– 8.8 8.4– 9.0	2 3 2 2	26 10 10½ 10½
12650	A 427	51.2 54.7	27 4 27	1919.68	220.4±1.7	1.62±0.02	8.8–13.8	2	40
13652	A 1240	30.2 23ʰ 33.7ᵐ	31 46 32° 9′	1916.64	350.2	1.93	9.0–12.0	1	40

BGC	Double Star	R. A. 1880 1950	Dec. 1880 1950	Epoch	Position Angle	Distance	Magnitudes	Nights Observed	Apt.
12655	Σ 3047 AB	23ʰ 51.8ᵐ 55.3	56°43′ 57 6	1916.69	71.6	1.05	9.0– 9.5	1	40
12655	β 280 AB & C			1916.69	196.1	8.31	. .–13.0	1	40
12664	β 730	52.5 56.1	— 4 13 — 3 50	1886.88	268.8±1.0	1.64±0.08	5.0– 9.6	3–2	26
12675	Σ 3050	53.4 56.9	33 4 27	1914.85	221.5±0.6	2.05±0.10	6.0– 6.4	4–3	10½
				1919.88	226.1±0.2	1.93±0.06	6.5– 6.9	3	10½
				1921.89	227.7±0.4	1.94±0.04	6.1– 6.4	3	10½
				1923.85	229.6±0.9	1.78±0.04	6.0– 6.2	2	10½
				1926.62	232.0±0.9	1.90±0.11	6.0– 6.1	3	10½
12677	β 731	53.4 57.0	— 8 28 5	1886.78	262.4±0.2	1.63±0.11	8.6– 9.4	2	26
				1888.84	262.9±1.1	1.35±0.08	8.2– 9.1	2	10
12700	β 482 AB	55.8 59.3	62 39 63 2	1916.88	342.7	4.24	9.0–10.0	1	10½
12700	β 482 AC			1916.88	124.1	10.33	. .–12.0	1	10½
12701	β 733 AB	55.9 59.5	26 27 50	1893.55		Not separated		1	16
				1914.70	120.5±1.4	0.86±0.04	6.0–11.0	4–2	40
				1916.63	141.0±2.9	0.85±0.04	6.0–10.0	2	40
12701	β 733 AC			1889.83	358.4±0.1	22.70±0.18	. .–. . .	4–2	10
				1914.68	335.6	52.08	. .–. . .	4–2	40
				1916.67	334.7±0.02	54.40±0.06	. .– 8.0	3	40
12701	β 733 AB & D			1916.68	295.1±0.1	104.52±0.50	. .–14.0	2	40
12709	β 281 AB	23 56.6 0 0.2	1 28 51	1888.89	210.1	1.35	8.0– 9.3	1	10
				1914.76	193.0	1.41	7.5–11.0	1	40
				1914.80	193.5±2.4	1.10±0.08	7.8–10.0	3	10½
				1915.92	191.2±0.4	1.46±0.04	7.3–10.3	3	10½
				1919.70	193.8	1.13	7.5–10.0	1	10½
				1922.77	192.5	1.18	7.5–10.5	1	10½
12709	β 281 AC			1914.76	333.9	34.40	. .–12.5	1	40
				1914.80	333.0±0.6	34.72±0.28	. .–11.8	3	10½
				1915.92	333.4±0.3	34.55±0.17	. .–11.2	3	10½
				1919.70	332.7	34.48	. .–11.3	1	10½
12723	A 429 AB	23 57.7 0 1.3	27 19 42	1914.74	342.2±2.0	0.53±0.03	8.9– 9.2	3	40
				1926.78	331.0	0.58	. .–. . .	1	10½
12723	h 1929 AB & C			1914.74	289.0±0.2	5.16±0.09	. .– 9.2	3	40
				1922.02	288.6±1.3	4.84±0.39	. .– 9.2	2	10½
		0ʰ 1.3ᵐ	27°42′	1926.78	287.5	4.96	. .– 9.5	1	10½

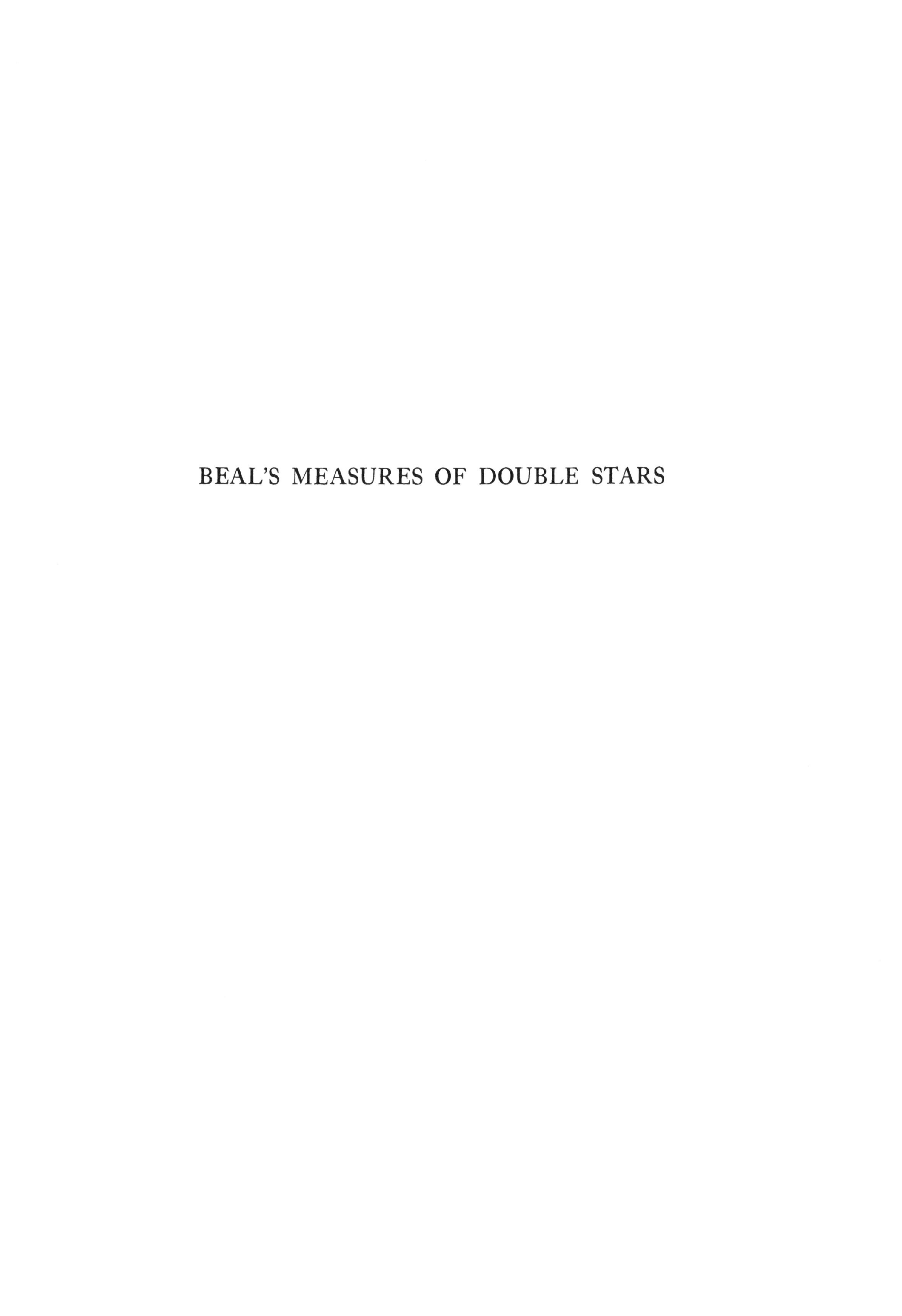

BEAL'S MEASURES OF DOUBLE STARS

WILLIAM OTIS BEAL
1874–1930

B.S. Earlham College, 1896

M.A. Haverford College, 1897

M.S. University of Chicago, 1902

Ph.D. (*magna cum laude*) University of Chicago, 1926

Instructor in Mathematics, Michigan State College, 1897–1900

Instructor in Mathematics, Chicago Manual Training High School, 1902–1903

Instructor in Mathematics, Illinois College, 1903–1904

Assistant Professor of Mathematics, Illinois College, 1904–1905

Professor of Mathematics and Physics, Illinois College, 1905–1912

Instructor in Astronomy, University of Minnesota, 1913–1926

Assistant Professor of Astronomy, University of Minnesota, 1926–1930

Member of the American Astronomical Society, Mathematical Society of America, Société Astronomique de France, and Sigma Xi.

MEASURES OF DOUBLE STARS

BGC	Double Star	R. A. 1880 1950	Dec. 1880 1950	Epoch	Position Angle	Distance	Magnitudes	Nights Observed	Apt.
52	β 254 AB	0ʰ 5.2ᵐ 8.9	59° 6′ 30	1915.89	235.6	7.60	7.6–10.6	4	10½
52	β 254 AC			1915.96	241.6	38.21	8.1–11.7	3	10½
106	β 392	10.5 14.2	60 52 61 16	1916.89	69.3	20.26	6.0–11.9	1	10½
247	β 107 AF	24.5 28.4	62 41 63 4	1917.84	113.8	151.96	10.3– 9.9	3	10½
247	β 107 AH			1917.84	92.5	45.39	10.3–11.8	3	10½
292	β 108 AB	27.7 31.6	62 15 38	1915.79 1917.91	359.3 360.9	4.44 3.77	7.4–11.6 7.6–10.3	3 4	10½ 10½
292	β 108 AF			1917.88	218.3	41.78	7.6–11.2	4	10½
364	β 109 AB	34.4 37.9	−17 10 −16 47	1917.93	354.5	104.97	8.2–10.7	3	10½
364	β 109 AC			1917.93	356.4	93.51	8.2–11.0	3	10½
395	β 231	38.0 42.9	47 38 48 1	1916.15	303.0	33.09	5.5–11.6	4	10½
426	Σ 60	41.7 45.7	57 11 34	1928.99	268.6	8.49	4.0– 7.6	1	10½
436	β 301 AC	43.4 46.9	−22 3 −21 40	1916.87 1917.86	297.2±1.1 298.6±0.2	10.86±0.37 10.50±0.32	8.3– 9.3 8.8–10.0	3 2	10½ 10½
440	β 232 AB & C	43.6 47.5	49 59 50 22	1915.92	294.2	27.79	8.1–10.1	3	10½
455	β 1 AB	45.8 49.8	55 58 56 21	1916.28	82.1	1.18	8.3–10.7	3	10½
455	β 1 AC			1916.07	135.5	3.66	8.2– 8.9	5	10½
455	β 1 AD			1916.08	194.6	9.06	8.2– 9.6	5	10½
455	β 1 AE			1916.22	331.5	15.53	8.3–11.7	4	10½
466	β 734	46.8 0 50.2	−24 40 17	1916.90 1917.92	346.7±0.3 346.8±0.2	11.41±0.43 11.84±0.53	6.4–11.0 6.5–11.0	3 2	10½ 10½
587	β 397	1 0.9 1ʰ 4.9ᵐ	46 12 46°35′	1916.91 1917.86	143.0±0.2 143.0±1.8	8.79±0.15 8.81±0.35	7.7–10.2 8.2–10.4	3 2	10½ 10½

BGC	Double Star	R. A. 1880 1950	Dec. 1880 1950	Epoch	Position Angle	Distance	Magnitudes	Nights Observed	Apt.
614	β 235 Aa	1ʰ 3.5ᵐ 7.6	50°22′ 44	1916.42	98.8	1.02	7.6– 7.8	4	10½
614	β 235 AB			1916.01	286.8	42.98	7.2–10.4	3	10½
614	β 235 AC			1916.01	67.8	60.57	7.2–10.0	4	10½
614	β 235 Cc			1916.23	48.3	7.73	10.1–11.2	4	10½
633	β 236	5.1 9.1	46 21 43	1915.80	113.5	5.18	8.3– 8.6	4	10½
662	β 3	9.6 13.9	55 52 56 14	1915.80	26.0	4.47	8.0–10.4	4	10½
758	β 82 AC	20.5 24.6	44 47 45 9	1916.88	110.7	122.26	5.2–10.9	3	10½
758	β 82 CD			1916.88	138.4	5.15	10.9–11.0	3	10½
923	β 736 AC	39.6 43.7	38 20 41	1916.92 1917.88	116.0±0.5 116.2±0.2	12.47±0.07 12.41±0.09	8.5– 8.8 8.7– 9.0	2 3	10½ 10½
990	β 259	46.3 49.8	−10 19 − 9 58	1915.92	235.9	4.54	8.5–10.8	3	10½
992	β 260	46.8 50.6	14 51 15 12	1916.38	244.8	1.03	8.3– 8.7	4	10½
995	β 183	47.4 1 50.8	−17 20 −16 59	1915.86	227.1	2.59	8.0– 9.3	4	10½
1409	β 261 AB	2 38.5 41.5	−28 25 7	1916.86	96.1	2.71	8.1–10.2	2	10½
1409	β 261 AC			1916.87	131.1	70.15	7.9–10.0	3	10½
1418	β 9	39.7 44.0	35 3 21	1916.88	172.0	1.90	6.1– 9.2	2	10½
1424	β 262	40.6 44.8	30 33 51	1916.08	60.9	1.63	8.1–10.5	5	10½
1427	Σ 305	40.7 44.6	18 52 19 10	1918.06	314.2±1.0	3.13±0.14	6.9– 7.8	5	10½
1460	β 10	44.4 2ʰ 47.9ᵐ	− 5 29 − 5°11′	1916.55	100.5	2.39	7.7–10.9	3	10½

BGC	Double Star	R. A. 1880 1950	Dec. 1880 1950	Epoch	Position Angle	Distance	Magnitudes	Nights Observed	Apt.
1512	Σ 333	2ʰ 52.4ᵐ 2 56.4	20°52′ 21 9	1918.06	203.7±1.1	1.28±0.05	6.0– 6.6	6–5	10½
1943	β 85	3 48.6 3 52.6	17 17 30	1915.99	214.4	4.33	8.0–10.5	3	10½
2159	β 744 AD	4 16.5 19.4	—26 1 —25 51	1916.88 1918.06	40.9 40.7±0.6	45.35 45.05±0.17	7.0– 8.0 6.8– 8.0	1 4	10½ 10½
2167	β 402 AB	17.0 20.5	— 1 33 23	1916.88 1918.16	74.5 73.5	7.38 7.62	9.2–11.5 9.2–11.6	1 1	10½ 10½
2287	β 88	31.6 4 35.1	— 2 43 34	1916.23	87.7	31.72	5.5–11.7	5	10½
2639	β 188 AB	5 11.8 15.2	— 6 58 53	1918.17	251.1	34.65	3.9–11.7	5	10½
2639	β 188 AD			1918.17	59.7	36.36	. . .–11.4	5	10½
2670	β 189	14.6 18.0	— 5 28 23	1916.13	283.8	4.22	7.5–11.9	3	10½
2673	β 190 AB & C	14.6 18.0	— 8 9 4	1916.08	4.2	35.11	8.1– 8.6	4	10½
2701	β 191	17.3 21.9	34 27 31	1916.11	23.9	3.71	10.2–10.3	3	10½
2857	β 1240 AB&C	30.9 35.4	30 25 28	1916.09	269.4	12.55	6.1– 8.7	3	10½
2857	β 90 AB & D			1916.09	112.9	33.23	6.1–11.4	3	10½
2902	Σ 774	34.7 38.2	— 2 0 — 1 57	1929.21	159.1±1.5	2.79±0.01	2.0– 5.0	2	10½
2931	A 118	37.4 41.3	13 16 18	1929.19	360.4	2.64	9.5– 9.6	1	10½
2968	β 192 AB	40.9 45.7	39 8 10	1916.19	353.3	39.67	5.3–11.6	3	10½
2968	β 192 AC			1916.19	34.2	49.64	5.3–10.9	3	10½
3008	β 94	44.2 47.4	—14 31 29	1916.10	175.9	2.76	6.1– 9.9	3	10½
. . .	AG	50.0 5ʰ 53.9ᵐ	10 32 10°33′	1918.19	192.6	22.60	9.1– 9.4	5	10½

BGC	Double Star	R. A. 1880 1950	Dec. 1880 1950	Epoch	Position Angle	Distance	Magnitudes	Nights Observed	Apt.
3271	β 96 AB	6ʰ10.5ᵐ 14.3	9°59′ 58	1916.25	257.4	63.05	6.2–11.1	3	10½
3271	β 96 AC			1916.25	159.2	119.14	6.2– 9.9	3	10½
3271	β 96 CD			1916.25	225.9	4.60	9.9–11.6	3	10½
3460	Σ 932	27.5 31.5	14 51 48	1929.21	323.0±1.0	1.94±0.04	8.2– 8.4	2	10½
3596	A. Clark 1	39.9 43.0	−16 33 37	1918.23	70.3±0.3	11.43±0.32	. . .– 9.9	3	10½
. . .	Espin 340	6 58.8 7 3.3	31 52 46	1918.22	138.5±0.5	5.15±0.12	9.2– 9.7	5	10½
3949	OΣ 170	11.1 14.9	9 31 24	1929.22	102.6±1.2	1.53±0.02	7.2– 7.5	3	10½
4060	β 198	20.6 23.6	−20 43 51	1917.22	211.1	5.86	8.0–11.1	3	10½
4122	Σ 1110 AB	27.0 31.5	32 9 0	1916.30 1918.20 1929.24	217.8±0.4 217.0±0.7 210.6±0.6	5.35±0.12 5.34±0.09 4.98±0.24	2.5– 3.1 2.2– 3.1	4 6 4	10½ 10½ 10½
4122	Σ 1110 AC			1918.20	164.4±0.2	73.25±0.16	2.0–10.0	5	10½
4164	β 200 AB	30.7 35.3	35 19 10	1916.28	190.6	100.92	5.5–10.8	3	10½
4164	β 200 AC			1916.27	99.3	161.40	5.5–10.0	4	10½
4164	β 200 CD			1916.34	241.3	1.67	10.6–11.5	2	10½
4192	β 201	33.7 36.8	−20 0 9	1916.12	332.8	2.95	7.9– 8.5	3	10½
. . .	Espin 904	35.3 40.7	51 29 20	1917.28 1918.28	127.5 131.6±0.8	3.64 2.86±0.24	9.3–10.8 9.5–11.0	1 3	10½ 10½
4269	Σ 1146	42.3 45.6	−11 54 −12 4	1929.22	9.0±0.4	3.46±0.11	5.3– 7.2	3	10½
4402	Σ 1175	56.1 59.8	4 29 18	1929.25	237.8	1.50	7.8– 8.7	1	10½
4405	β 23	56.2 7ʰ59.9ᵐ	3 26 3°15′	1917.23	180.6	2.79	8.3–11.8	3	10½

BGC	Double Star	R. A. 1880 1950	Dec. 1880 1950	Epoch	Position Angle	Distance	Magnitudes	Nights Observed	Apt.
4413	β 203 AB	7ʰ 57.7ᵐ 8 0.6	−27°13′ 25	1916.12 1918.18	243.7 244.7	7.04 6.84	7.7– 8.7 7.6– 9.6	3 5	10½ 10½
4413	β 203 AC			1918.18	74.2	63.56	. . .– 9.5	5	10½
4477	Σ 1196 AB	5.3 9.3	18 1 17 49	1918.36 1929.24	291.4±0.2 105.0±0.7	0.76±0.09 5.46±0.06	5.2– 5.5 5.0– 5.7	3 4	10½ 10½
4477	Σ 1196 AC			1918.36	112.0±0.2	5.08±0.05	5.2– 6.2	3	10½
. . .	Espin 592	4.9 9.7	41 55 43	1917.26	327.8±0.11	3.09±0.04	8.5–10.6	3	10½
. . .	Espin 426	17.6 21.9	28 51 38	1917.31	284.7±1.5	4.37±0.01	9.3– 9.7	3	10½
4708	β 207	33.3 36.5	−19 19 33	1916.18	100.1	4.24	6.2–10.7	3	10½
4714	β 208 AB	33.9 37.0	−22 16 31	1918.27	195.7	1.31	6.0– 8.4	3	10½
4714	β 208 AC			1918.23	201.8	84.98	6.0–10.7	5	10½
4723	Σ 1258	34.9 39.9	49 18 3	1918.35	331.1±0.2	9.88±0.04	7.1– 7.5	4	10½
4730	β 209	35.4 40.0	39 14 38 59	1916.44	1.1	1.25	8.5– 8.6	4	10½
4771	Σ 1273	40.4 44.1	6 52 37	1929.25	252.7	3.56	3.0– 7.2	1	10½
. . .	Espin 596	39.7 44.5	45 58 43	1917.26	201.6±0.1	3.21±0.03	8.6–10.5	3	10½
4839	Σ 1291	46.9 51.2	31 2 30 46	1929.28	321.4±0.6	1.62±0.18	6.0– 6.0	2	10½
4849	β 24	48.4 51.8	− 8 18 34	1916.21	177.3	1.05	7.7– 8.8	3	10½
4853	β 103	49.0 8 52.4	− 7 22 38	1916.96	73.2	2.80	8.1–11.6	4	10½
4972	Σ 1321	9 6.4 9ʰ 11.0ᵐ	53 13 52°56′	1918.35	69.4±0.2	19.07±0.12	6.7– 7.0	4	10½

BGC	Double Star	R. A. 1880 1950	Dec. 1880 1950	Epoch	Position Angle	Distance	Magnitudes	Nights Observed	Apt.
5030	Σ 1338	9ʰ 13.5ᵐ 17.9	38°42′ 25	1918.36	179.5±0.6	1.34±0.10	6.6– 7.1	4	10½
. . .	Espin 718	15.1 20.0	52 7 51 49	1917.30 1918.28	28.0 27.5±0.8	7.57 7.67±0.23	9.3–10.5 9.3–11.4	1 4	10½ 10½
5062	β 105 AB	17.7 21.8	26 42 24	1916.22 1918.27	205.6 206.6	2.91 3.06	5.3–10.8 4.8–10.9	3 4	10½ 10½
5062	β 105 AC			1918.26	211.3	153.07	5.0–10.0	5	10½
5071	Σ 1348	18.2 21.9	6 52 34	1918.31 1929.28	320.6±1.8 321.0±1.6	1.67±0.13 2.13±0.13	7.5– 7.7 7.0– 7.0	5 2	10½ 10½
5094	Σ 1355	21.0 24.7	6 46 28	1918.32 1929.28	337.2±0.7 338.5±0.7	2.43±0.10 2.71±0.31	7.5– 7.7 7.2– 7.2	4 2	10½ 10½
5181	β 214	35.9 39.2	−17 56 −18 15	1917.22	256.5	3.03	7.5–11.2	3	10½
5263	β 216	51.3 9 54.5	−25 59 −26 19	1918.30	159.4	3.19	6.2–11.6	2	10½
5325	β 217	10 1.3 4.5	−24 18 38	1916.48	287.1	1.58	8.2– 8.3	4	10½
5329	β 218	1.7 5.0	−19 7 27	1916.27	129.1	0.86	8.5– 8.9	3	10½
. . .	Espin 180	1.5 5.6	36 10 35 50	1917.27	358.1±0.9	8.03±0.55	9.0–10.8	4	10½
5365	OΣ 215	9.7 13.5	18 20 17 59	1917.39	201.6±0.5	1.07±0.11	6.9– 7.4	3	10½
5388	Σ 1424	13.3 17.1	20 27 6	1918.26 1929.28	117.0±1.1 116.9±0.2	3.96±0.12 4.31±0.02	2.4– 4.0 2.0– 3.0	5 2	10½ 10½
. . .	Espin 432	19.0 23.0	33 14 32 53	1917.37 1918.27	161.7 163.5±0.3	3.03 3.19±0.24	9.5– 9.5 9.5–10.3	1 4	10½ 10½
. . .	Espin 1151	23.6 27.8	44 52 31	1917.34 1918.30	299.4 301.4±0.8	2.65 2.37±0.07	9.5– 9.7 9.6–10.5	1 4	10½ 10½
. . .	Jon 736	24.0 27.8	15 21 0	1918.31	202.2±1.2	2.38±0.15	10.2–10.6	4	10½
. . .	A 1350	25.3 10ʰ 28.9ᵐ	− 1 49 − 2°10′	1918.30	324.7±1.1	2.45±0.14	9.5–10.3	5	10½

BGC	Double Star	R. A. 1880 1950	Dec. 1880 1950	Epoch	Position Angle	Distance	Magnitudes	Nights Observed	Apt.
5508	Σ 1457	10ʰ 32.5ᵐ	6°21′	1917.41	321.2±2.5	1.40±0.09	7.4– 8.0	5	10½
		36.1	5 59	1918.26	324.5±1.0	1.32±0.06	7.6– 8.7	5	10½
5560	OΣ 229	41.1	41 46	1917.42	312.3±2.2	0.82±0.08	7.1– 7.4	5	10½
		45.1	24						
5579	β 111	45.2	– 8 28	1916.21	4.8	3.60	9.7– 9.8	3	10½
		48.7	50						
. . .	Jon 427	10 57.5	16 42	1917.34	102.3±1.1	4.34±0.10	9.9–10.2	2	10½
		11 1.2	19	1918.27	102.5±0.8	4.24±0.33	9.9–10.7	3	10½
. . .	Espin 181	6.6	36 22	1917.30	143.0	5.63	9.0–10.7	1	10½
		10.5	35 59	1918.32	142.2±0.5	5.48±0.29	9.5–11.1	4	10½
. . .	Espin 305	10.9	35 8	1917.34	32.8±1.0	4.33±0.36	8.9– 9.9	3	10½
		14.7	34 45						
5765	Σ 1536	17.6	11 12	1918.30	39.2±1.9	1.84±0.16	4.6– 7.8	5	10½
		21.2	10 49						
5766	β 26	17.7	– 9 46	1916.24	68.2	2.53	8.0–10.9	3	10½
		21.2	–10 9						
. . .	Espin 1401	28.1	41 31	1917.30	328.7	6.95	10.3–10.6	1	10½
		11 31.9	8	1918.34	328.7	7.06	10.3–11.2	1	10½
6053	Σ 1606	12 4.7	40 34	1917.43	324.8±1.4	0.94±0.05	6.5– 7.1	5	10½
		8.2	10						
6243	Σ 1670	35.6	– 0 47	1914.35	325.2	6.08	. .–. . .	1	10½
		39.2	– 1 10	1916.28	324.4±0.3	6.04±0.05	3.0– 3.0	3	10½
				1917.42	323.1±0.8	6.02±0.10	3.1– 3.2	4	10½
				1918.36	323.4±0.4	6.15±0.06	3.2– 3.4	2	10½
. . .	Jon 1014	47.0	7 18	1917.41	61.7±1.0	3.78±0.18	9.2– 9.8	2	10½
		50.5	6 55						
6295	Σ 1686	47.0	15 41	1914.38	186.8±0.8	5.67±0.17	. .–. . .	2	10½
		50.5	18	1916.28	187.8±0.3	5.77±0.18	8.3– 8.6	3	10½
6345	β 112 AB	54.8	19 1	1916.28	351.1	149.16	6.3– 9.6	3	10½
		58.3	18 38						
6345	β 112 BC	12 58.3	18 38	1916.28	292.6	1.82	9.6–10.1	3	10½
. . .	Jon 434	13 2.4	– 0 13	1917.43	325.9±2.0	3.63±0.12	9.2– 9.5	3	10½
		13ʰ 6.0ᵐ	– 0°36′						

BGC	Double Star	R. A. 1880 1950	Dec. 1880 1950	Epoch	Position Angle	Distance	Magnitudes	Nights Observed	Apt.
6415	OΣ 261	13ʰ 6.4ᵐ	32°43′	1914.68	345.8	1.76	. .–. . .	1	10½
		9.7	21	1916.33	343.7±1.0	1.71±0.08	6.9– 7.4	3	10½
				1917.45	344.5±1.8	1.63±0.14	6.5– 6.8	5	10½
6490	β 237	21.0	15 0	1916.29	205.8	2.53	8.3–10.8	4	10½
		24.4	14 38						
6528	β 114	28.0	— 8 0	1916.28	142.6	1.26	7.9– 8.0	3	10½
		31.7	22						
6566	Σ 1768	32.1	36 54	1917.51	126.2±1.1	1.26±0.06	5.6– 8.6	5	10½
		35.2	32	1928.57	123.7±0.8	1.56±0.07	5.7– 7.6	4	10½
. . .	Espin 309	38.5	32 10	1917.37	135.0	1.95	9.4– 9.7	1	10½
		41.7	31 49						
6626	Σ 1783	40.9	41 38	1914.37	50.6	2.40	. .–. . .	1	10½
		43.9	17	1916.33	48.9±1.4	1.79±0.16	8.0–11.0	3	10½
6641	Σ 1785	43.6	27 35	1917.45	354.0±1.3	1.26±0.04	6.9– 7.3	5	10½
		46.8	14	1928.54	61.3±0.6	1.20±0.08	7.2– 7.5	4	10½
6690	β 30	52.4	20 3	1916.34	197.7	8.05	8.3–11.2	3	10½
		13 55.7	19 42						
6811	β 116	14 13.0	—13 9	1916.34	277.7	3.18	8.1– 8.5	3	10½
		16.8	29						
. . .	Jon 439	16.6	1 13	1917.37	237.1	4.47	9.4–10.5	1	10½
		20.2	0 54						
6857	β 225 AB	18.8	—19 26	1916.36	295.6	35.30	6.8– 7.4	4	10½
		21.7	45						
6857	β 225 BC			1916.36	98.2	1.21	7.5– 8.5	3	10½
6912	β 238	27.0	—20 30	1916.34	91.0	6.84	8.5–11.1	3	10½
		30.9	49						
6941	β 226	32.1	—21 49	1916.38	94.9	1.04	8.2– 8.6	3	10½
		36.1	—22 7						
7034	Σ 1888	45.8	19 36	1928.54	31.8±0.5	3.94±0.08	4.7– 6.6	4	10½
		49.0	15						
7041	β 118	47.0	—16 1	1917.52	301.2	1.70	8.9–10.0	4	10½
		50.9	18						
7070	β 239	51.6	—27 10	1917.50	324.5	0.81	6.1– 6.5	3	10½
		14ʰ 55.7ᵐ	—27°27′						

BGC	Double Star	R. A. 1880 1950	Dec. 1880 1950	Epoch	Position Angle	Distance	Magnitudes	Nights Observed	Apt.
7120	Σ 1909	14h 59.8m 15 2.2	48° 7′ 24	1928.55	246.2±0.3	3.35±0.09	5.2– 6.1	4	10½
7195	β 352	10.7 14.9	—26 33 49	1917.49	66.2±0.2	14.53±0.11	8.4–10.1	3	10½
7208	β 228	12.6 16.7	—23 50 —24 3	1916.53	319.3	0.96	8.3– 9.0	2	10½
7211	Lv 6	13.0 17.2	—26 35 51	1917.49	29.4±0.2	17.27±0.04	8.4–10.5	3	10½
7259	Σ 1938	20.0 22.7	37 46 31	1928.58	41.1±0.6	1.61±0.10	6.7– 7.3	4	10½
7336	β 121	32.3 36.5	—27 15 29	1916.48	277.7	1.43	8.3– 8.5	3	10½
7340	β 122	33.0 37.0	—19 23 37	1916.36	211.2	1.88	7.6– 7.9	3	10½
7380	β 240 AB	39.5 43.0	4 24 10	1916.66	134.5	1.77	8.5–10.5	5	10½
7380	β 240 AC			1916.71	38.4	29.33	8.5–11.8	4	10½
7418	β 36	46.4 50.2	—24 58 —25 11	1916.48	276.7	2.47	6.1– 8.7	3	10½
7476	β 37	55.2 59.4	—24 15 27	1916.48	44.0	3.03	8.8–10.3	3	10½
7478	β 38	55.6 59.8	—24 41 53	1916.76	351.4	5.02	8.3–10.6	4	10½
. . .	AG Camb 7446	15 57.2 16 0.1	27 27 15	1917.50	226.4±0.8	12.09±0.15	8.6–10.1	3	10½
7487	Σ 1998 AB	15 57.8 16 1.6	—11 3 15	1917.57	174.9±0.8	0.87±0.03	5.1– 5.2	5	10½
7487	Σ 1998 AB & C			1917.57	59.6±0.5	7.57±0.20	5.2– 9.1	5	10½
7502	β 39	1.0 4.9	—12 25 37	1916.51	258.8	3.21	6.5–10.9	3	10½
7530	β 40	4.5 16h 8.8m	—27 14 —27°25′	1916.49	356.0	5.37	8.5–10.5	3	10½

BGC	Double Star	R. A. 1880 1950	Dec. 1880 1950	Epoch	Position Angle	Distance	Magnitudes	Nights Observed	Apt.
7533	β 120 AB	16^h 5.0^m 9.1	—19° 9′ 20	1916.52	6.5	0.96	4.5– 6.5	3	10½
7533	β 120 AB & C			1916.51	336.6	41.42	4.5– 7.9	4	10½
7533	β 120 CD			1916.51	50.7	2.11	7.9– 9.1	4	10½
7563	Σ 2032 AB	10.2 12.8	34 10 33 59	1917.49 1928.60	218.0±0.3 221.8±0.2	5.25±0.16 5.39±0.05	5.3– 6.4 5.0– 6.1	5 8	10½ 10½
7563	Σ 2032 AD			1917.49	85.3±0.3	66.82±0.39	5.6– 9.3	5	10½
7603	β 41	17.4 18.4	61 44 34	1916.65	54.2	1.88	8.7–10.6	4	10½
7712	β 42	35.3 38.1	29 15 7	1916.52	41.3	7.62	9.5–10.0	5	10½
7786	β 123	47.5 51.7	—21 51 58	1916.60	201.3	1.80	8.5– 9.1	3	10½
7791	Σ 241	48.4 16 52.5	—21 22 29	1916.58	353.3	0.78	7.5– 7.5	2	10½
7878	Σ 2130	17 2.8 4.3	54 38 32	1928.75	115.1±0.7	2.62±0.08	5.0– 5.1	4	10½
7887	β 124	4.0 7.6	— 0 36 42	1917.26	263.5	0.88	7.6–10.1	4	10½
7922	Σ 3127	10.1 13.0	24 59 54	1928.63	207.5±0.4	10.98±0.08	3.0– 7.5	4	10½
7951	β 127	13.4 17.8	—27 13 17	1916.46	93.6	5.10	8.9–10.1	3	10½
8000	β 128	19.4 23.7	—26 14 18	1916.56	322.1	3.87	8.0–11.0	3	10½
8014	β 129	21.2 25.5	—25 24 28	1917.20	103.6	0.97	7.5– 8.2	3	10½
8038	Σ 2173	24.2 27.8	— 0 58 — 1 2	1917.52	157.3±1.2	0.83±0.10	5.9– 6.2	5	10½
8288	β 47	54.9 58.8	—10 14 15	1916.92	276.4	1.11	8.6–10.9	3	10½
8303	Σ 2262	17 56.6 18^h 0.4^m	— 8 11 — 8°11′	1917.51	262.3±1.0	2.03±0.06	5.2– 5.7	5	10½

BGC	Double Star	R. A. 1880 1950	Dec. 1880 1950	Epoch	Position Angle	Distance	Magnitudes	Nights Observed	Apt.
8340	Σ 2272	17ʰ 59.4ᵐ 18 2.9	2°33′ 33	1917.59 1928.60	136.9±0.6 124.6±0.4	5.06±0.12 6.43±0.10	4.3– 7.2 4.3– 6.2	5 8	10½ 10½
8355	β 243 AB	0.9 5.1	−22 17 17	1917.51	121.4	0.64	8.8– 9.7	1	10½
8355	β 243 AB & C			1916.52	56.8	40.71	8.3– 9.3	3	10½
8356	β 244	1.0 5.4	−27 53 53	1917.49	257.6	2.24	8.2–10.2	2	10½
8371	β 245	2.4 6.9	−30 45 45	1917.38	354.0	4.65	6.3– 9.6	5	10½
8414	β 131 AB	6.7 10.7	−15 38 37	1916.54	278.1	2.72	7.5–10.5	3	10½
8488	β 48	13.9 18.0	−19 43 42	1917.36	360.1	2.07	8.2–10.8	4	10½
8520	β 49 AB	17.0 21.1	−19 38 36	1916.59	46.9	8.06	8.3–10.9	4	10½
8554	A 580	20.7 24.1	7 37 39	1928.60	321.7±1.2	4.38±0.40	8.7–10.8	4	10½
8549	β 133	20.2 24.6	−26 42 40	1916.49	260.7	1.29	7.5– 7.8	3	10½
8617	β 247	25.6 29.4	− 9 27 24	1916.58	168.3	7.68	8.3–11.2	3	10½
8619	β 419	25.7 29.5	− 7 55 52	1916.65	44.3±2.4	1.14±0.02	8.3– 9.3	3	10½
8663	OΣ 358	30.5 33.6	16 53 56	1917.59 1928.59	187.6±0.2 183.2±0.7	1.84±0.06 2.10±0.13	7.4– 7.6 6.8– 7.2	5 4	10½ 10½
8710	β 50 AC	34.2 36.5	39 29 32	1916.83	329.0	73.81	8.8–10.3	3	10½
8776	Σ 2375	39.6 43.0	5 22 25	1928.68	115.4±0.7	2.38±0.06	6.6– 7.0	4	10½
8798	Σ 2398	41.6 18ʰ 42.7ᵐ	59 25 59°29′	1928.75	155.5±0.5	17.01±0.16	8.2– 8.7	4	10½

BGC	Double Star	R. A. 1880 1950	Dec. 1880 1950	Epoch	Position Angle	Distance	Magnitudes	Nights Observed	Apt.
8804	β 51 AB	18ʰ 41.7ᵐ / 18 44.0	39°34′ / 38	1916.75	186.0	75.03	8.5–10.7	3	10½
8804	β 51 BC			1916.75	297.2	6.49	10.7–11.5	3	10½
9043	Σ 2455	19 1.8 / 4.8	21 59 / 22 5	1928.64	56.4±0.2	4.81±0.10	7.3– 8.2	4	10½
9116	β 139 CD	7.2 / 10.3	16 39 / 46	1917.93	261.4	127.55	7.9–10.1	1	10½
9154	β 140 AB	10.2 / 14.1	—11 11 / 4	1916.58	321.4	39.42	7.6–10.5	4	10½
9154	β 140 BC			1916.61	209.0	7.99	10.6–11.5	3	10½
9313	β 142	21.5 / 25.4	—12 23 / 15	1916.55	348.4	1.55	8.1– 8.2	4	10½
9319	Σ 2525	22.7 / 25.5	27 5 / 13	1928.74	302.4±0.8	1.23±0.02	7.5– 7.7	4	10½
9387	β 143	26.6 / 28.5	49 15 / 24	1916.72	191.8	1.84	8.4– 9.8	3	10½
9424	β 53	29.8 / 33.1	11 11 / 20	1916.86	250.9	1.52	8.8–10.2	3	10½
9507	Σ 2557 AB	34.8 / 37.6	29 28 / 37	1916.74	104.3	11.36	8.2– 9.8	2	10½
9507	Σ 2557 AC			1916.74	304.2	21.73	8.2–11.5	2	10½
9594	β 55 AB	40.5 / 43.8	10 16 / 26	1915.83	26.1	4.29	10.0–10.0	3	10½
9623	β 147	42.3 / 45.0	31 48 / 58	1916.74	300.0	8.80	8.9–10.5	3	10½
9669	A 376	45.9 / 49.3	7 20 / 30	1928.68	126.6±2.1	2.00±0.12	9.0–10.0	4	10½
9693	OΣ 388	47.3 / 50.3	25 33 / 43	1928.67	138.7±0.5	3.77±0.11	7.2– 7.4	4	10½
9712	Σ 2596	48.5 / 51.7	14 59 / 15 10	1928.68	322.6±1.1	2.02±0.07	7.2– 8.4	4	10½
9868	Σ 2624	19 59.0 / 20 1.6	35 41 / 52	1928.75	174.9±0.5	2.03±0.12	6.7– 7.2	4	10½
9924	β 58	1.8 / 20ʰ 5.0ᵐ	15 44 / 15°56′	1915.77	187.2	9.20	7.2–10.3	3	10½

BGC	Double Star	R. A. 1880 1950	Dec. 1880 1950	Epoch	Position Angle	Distance	Magnitudes	Nights Observed	Apt.
9926	Σ 2628	20ʰ 2.0ᵐ 5.4	9° 3′ 15	1928.74	344.0±0.4	4.04±0.15	6.1– 8.0	4	10½
. . .	AG Camb 10992	7.4 10.2	29 1 13	1917.81	304.3±0.7	3.71±0.12	8.5– 9.0	3	10½
10047	β 59	10.6 14.1	4 45 58	1917.81	114.2	8.39	9.3–10.5	3	10½
10427	β 267	35.4 39.1	— 4 49 34	1916.80	239.5	1.75	9.6– 9.7	3	10½
10480	A171	39.1 42.2	20 53 21 8	1928.82	324.4±1.1	5.14±0.16	8.2–11.5	4	10½
10489	A 172	39.5 42.6	20 35 50	1928.84	217.9±1.0	2.94±0.20	9.0–10.5	4	10½
10538	β 66	43.0 46.0	27 1 16	1916.81	162.5	1.18	8.3– 8.9	3	10½
10588	β 155 AB	47.4 49.5	50 58 51 14	1917.33	27.0	0.88	7.3– 8.0	2	10½
10605	Σ 2735	49.7 53.2	4 4 20	1928.80	285.4±0.7	1.97±0.16	6.0– 7.5	4	10½
10643	Σ 2737 AB & C	53.1 56.6	3 50 4 6	1928.80	72.2±0.1	10.81±0.15	5.7– 7.0	4	10½
10676	Σ 2742	56.3 59.8	6 43 59	1928.75	219.3±0.6	2.89±0.04	6.9– 7.1	4	10½
10685	Σ 2744	20 57.0 21 0.6	1 4 20	1928.75	152.5±1.0	1.59±0.04	6.0– 7.0	4	10½
10692	Lv 9	20 57.4 21 0.1	38 45 39 1	1916.85 1917.88	194.6±2.5 190.3	2.28±0.14 1.63	9.6–10.8 9.5–11.2	3 1	10½ 10½
10732	Σ 2758	1.2 3.9	38 8 25	1915.79 1928.64	130.2±0.2 133.9±0.4	23.76±0.12 25.08±0.19	5.3– 5.9 5.3– 5.9	6 4	10½ 10½
10746	Σ 2760	1.9 4.8	33 39 56	1928.75	230.2±0.6	3.60±0.11	7.0– 8.0	4	10½
10782	β 71 AB	4.5 7.9	9 39 56	1916.89	275.7	2.42	4.5–10.8	3	10½
10782	β 71 AB & C			1916.83	6.4	47.21	4.7–11.5	4	10½
10782	β 71 AB & D	21ʰ 7.9ᵐ	9°56′	1916.83	152.5	354.03	4.7– 6.0	4	10½

BGC	Double Star	R. A. 1880 1950	Dec. 1880 1950	Epoch	Position Angle	Distance	Magnitudes	Nights Observed	Apt.
10808	β 159 AB & C	21ʰ 6.4ᵐ 8.8	47°12′ 29	1916.03	189.4	135.64	6.3– 7.2	5	10½
10824	β 160 AB	7.8 10.3	45 13 30	1916.75	154.6	57.44	7.7–10.8	3	10½
10824	β160 BC			1916.75	114.7	6.04	10.8–11.0	3	10½
10884	β 252	13.0 17.1	−27 49 32	1916.84	272.7	2.40	8.4– 8.6	3	10½
10997	β 369 AB	22.5 24.8	52 14 32	1916.87 1917.88	30.8±1.1 29.7	16.60±0.48 17.15	8.0–11.4 7.8–11.6	3 1	10½ 10½
10997	β 369 AC			1916.86 1917.88	62.6±0.6 62.7	50.89±0.21 52.07	8.0–11.2 7.8–11.4	3 1	10½ 10½
11001	Σ 2799	23.0 26.4	10 34 52	1928.75	287.2±0.2	1.48±0.03	6.5– 6.5	4	10½
11026	β 73 AB	25.2 28.9	− 6 6 − 5 48	1915.87	318.8	35.66	3.0–10.1	4	10½
11026	β 73 AC			1915.90	185.8	57.20	. .–11.3	3	10½
11056	β 165	27.9 31.5	− 3 59 41	1915.82	178.8	5.31	8.5–10.5	3	10½
11058	β 370	28.2 30.5	52 13 31	1916.90 1917.85	326.5±0.8 325.6±0.0	3.52±0.08 3.14±0.13	8.5– 9.4 8.4– 9.5	3 2	10½ 10½
11076	β 166	30.3 32.2	59 48 60 7	1916.80	261.2	0.96	. .–. . .	2	10½
11121	β 371	33.0 35.1	58 10 29	1916.86 1917.86	2.5 2.1±1.1	8.53 8.63±0.16	8.0–10.6 8.4–10.8	1 3	10½ 10½
11178	β 274	36.4 39.2	38 56 39 15	1916.76	178.3	3.03	8.3–11.2	3	10½
11214	Σ 2822	38.8 21 41.9	28 12 31	1928.86	156.2±0.2	1.10±0.09	5.0– 6.5	4	10½
11483	Σ 2863	22 0.3 2.3	64 2 22	1928.87	279.3±0.4	7.32±0.12	5.0– 7.0	4	10½
11490	Σ 2862	1.0 22ʰ 4.6ᵐ	− 0 1 + 0°19′	1928.76	99.7±0.4	2.39±0.12	7.4– 7.8	4	10½

BGC	Double Star	R. A. 1880 1950	Dec. 1880 1950	Epoch	Position Angle	Distance	Magnitudes	Nights Observed	Apt.
11590	Σ 2878	22ʰ 8.5ᵐ 12.0	7°23′ 44	1928.85	123.4±1.0	1.51±0.04	7.2– 8.5	4	10½
11599	Σ 2881	9.1 12.3	28 59 29 20	1928.78	95.1±0.6	1.55±0.17	7.4– 7.8	4	10½
11691	β 172 AB & C	17.9 21.5	— 5 27 6	1917.82	341.8	55.58	6.5–11.6	3	10½
11691	β 172 AB & D			1917.82	191.0	115.96	6.5–10.6	3	10½
11691	β 172 AB & E			1917.82	133.3	132.43	6.5– 9.8	3	10½
11696	Σ 2903	18.2 20.3	66 6 27	1928.90	94.6±0.3	4.32±0.09	6.9– 8.0	4	10½
11735	β 380 AB	22.0 24.8	49 6 27	1916.94 1917.86	323.2 323.0±0.6	25.51 24.85±0.22	7.3–11.1 8.6–11.4	1 3	10½ 10½
11735	β 380 AC			1916.92 1917.86	134.5±0.0 134.3±0.1	36.28±0.03 36.49±0.08	7.2– 7.6 8.6– 8.5	2 3	10½ 10½
11738	β 173	22.4 25.0	56 35 56	1916.83	232.0	2.62	8.3–11.0	3	10½
11743	Σ 2909	22.6 26.2	— 0 38 17	1917.87 1928.76	305.4±0.3 301.2±0.6	2.91±0.05 2.82±0.06	3.9– 4.1 4.0– 4.1	5 4	10½ 10½
11752	β 478 AC	23.1 26.8	— 7 56 35	1916.93 1917.89	238.7 237.3±0.5	29.53 29.21±0.18	8.8– 9.2 9.0– 9.6	1 2	10½ 10½
11832	β 175	29.8 31.3	74 24 46	1916.55	315.8	1.62	10.1–10.3	3	10½
11866	Σ 2928	33.2 36.9	—13 14 —12 52	1914.84 1915.79 1917.87	127.0 127.6±0.4 127.6	4.18 3.80±0.06 3.92	. .–. . . 8.0– 8.1 8.5– 8.8	1 3 1	10½ 10½ 10½
11903	β 709	35.4 39.0	— 3 11 — 2 49	1928.87	5.0±0.6	2.22±0.12	8.0– 8.7	4	10½
11914	Σ 2935	36.8 40.5	— 8 56 34	1928.86	307.6±1.0	2.69±0.17	6.4– 7.5	4	10½
11968	Σ 2944	41.7 45.3	— 4 51 29	1928.80	265.2±1.0	3.01±0.13	7.0– 7.2	4	10½
11991	Σ 2946	44.2 22ʰ47.4ᵐ	39 53 40°15′	1928.91	258.3±0.3	5.41±0.15	7.9– 8.0	4	10½

BGC	Double Star	R. A. 1880 1950	Dec. 1880 1950	Epoch	Position Angle	Distance	Magnitudes	Nights Observed	Apt.
11997	Σ 2947	22ʰ 44.9ᵐ 47.2	67°56′ 68 18	1928.89	61.6±1.0	3.97±0.08	6.9– 7.1	4	10½
12012	β 177	45.9 49.7	−22 21 −21 59	1915.78	97.6	2.62	7.8– 7.8	3	10½
12021	Σ 2950	46.7 49.4	61 3 61 25	1928.90	302.0±0.4	2.30±0.12	6.0– 7.2	4	10½
12065	Σ 2958	50.9 54.4	11 12 34	1928.88	11.8±0.3	4.06±0.28	6.7– 8.3	4	10½
12094	OΣ 483	53.2 22 56.7	11 5 27	1928.86	249.0±2.6	1.08±0.04	6.0– 7.6	2	10½
12172	Σ 2976 AB	23 1.6 5.1	5 57 6 20	1928.88	263.2±0.8	7.60±0.10	8.3–10.2	4	10½
12172	Σ 2976 AC			1928.88	196.1±0.2	18.41±0.14	. .– 8.8	4	10½
12176	β 78 AB	2.2 5.5	30 49 31 12	1915.82	55.2	18.77	7.2–10.8	3	10½
12176	β 78 AC			1915.82	61.9	48.86	7.2–10.9	3	10½
12177	β 180 AB	2.2 5.1	60 11 34	1915.91	170.5	0.81	7.5– 8.4	3	10½
12177	β 180 AB & C			1915.84	106.2	34.80	7.5–10.2	3	10½
12201	β 385 AB & C	4.5 7.8	31 50 32 13	1916.89 1917.84	76.9±0.0 77.0±0.2	58.39±0.10 58.36±0.08	7.3– 8.9 7.2– 9.4	3 2	10½ 10½
12290	β 80 AB & C	12.8 16.4	4 45 5 8	1917.84	359.6	105.72	8.4–10.9	3	10½
12290	β 80 AB & D			1917.84	330.4	192.71	8.4– 9.4	3	10½
12290	β 80 AB & E			1917.84	294.8	206.07	8.4– 9.7	3	10½
12293	A 202	12.8 16.0	46 36 59	1928.92	257.4±0.7	2.83±0.10	7.7–10.2	2	10½
12308	β 229	14.4 17.5	56 35 58	1915.79	35.9	17.42	7.0–11.7	3	10½
12316	β 278	15.3 23ʰ 18.3ᵐ	61 33 61°58′	1915.88	175.1	12.93	7.0–11.6	3	10½

BGC	Double Star	R. A. 1880 1950	Dec. 1880 1950	Epoch	Position Angle	Distance	Magnitudes	Nights Observed	Apt.
12332	Σ 3007	23ʰ 16.8ᵐ 20.3	19°54′ 20 17	1928.88	89.6±0.5	6.46±0.30	7.0–10.3	4	10½
12340	Σ 3008	17.5 21.1	— 9 7 — 8 44	1928.90	208.7±0.9′	3.46±0.17	7.0– 8.0	4	10½
. . .	AG Leip I 9311	20.6 24.1	14 29 52	1917.92 1918.00	252.5±0.3 250.2	7.38±0.42 7.74	9.0–11.6 9.0–11.3	2 1	10½ 10½
12435	β 387	28.1 31.7	—10 22 — 9 59	1916.91 1917.87	71.1±1.1 70.9±2.0	5.75±0.25 6.25±0.45	8.2– 9.6 8.3–10.3	3 2	10½ 10½
12517	OΣ 503	36.0 39.5	19 38 20 1	1914.79 1915.80 1917.87	134.3 130.5±0.8 128.2±0.8	1.59 1.53±0.04 1.25±0.09	. .–. . . 7.2– 7.9 7.8– 8.5	1 3 2	10½ 10½ 10½
12655	Σ 3047 AB	51.8 55.3	56 43 57 7	1916.80	71.1	1.13	8.4– 8.8	3	10½
12655	β 280 AC			1916.55	191.1	8.37	8.4–11.9	3	10½
12675	Σ 3050	53.4 23ʰ 57.0ᵐ	33 4 33°28′	1928.86	233.5	2.08±0.06	6.0– 6.1	4	10½

PUBLICATIONS OF
FRANCIS P. LEAVENWORTH

1883   Observations of Comets of 1880, 1881, and 1882.
Publications of Cincinnati Observatory, No. 7.

1886   Observations of Nebulae Supposed to Be New.
Astronomical Journal, No. 146, 7:9–14, November 24.

1887   Lalande 4219.
Sidereal Messenger, 6:80–81, February.

Observations of Nebulae Supposed to Be New.
Astronomical Journal, No. 152, 7:57–61, February 17.

Observations of New Double Stars.
Astronomical Journal, No. 156, 7:95, May 20.

Observations of Sappho (80).
Astronomical Journal, No. 157, 7:101, June 6.

Observations of Occultations.
Astronomical Journal, No. 158, 7:112, July 6.

Note on N.G.C. 4333.
Sidereal Messenger, 6:293–294, September.

1888   Double Stars h 3823 and h 4321.
Sidereal Messenger, 7:172, April.

Filar-Micrometer Observations of Comet 1888 I.
Astronomical Journal, No. 173, 8:39, June 4.

Filar-Micrometer Observations of Comet 1888 I.
Astronomical Journal, No. 176, 8:63, July 12.

Comet a 1888 (Sawerthal).
Sidereal Messenger, 7: pp. 308–309, August.

Filar-Micrometer Observations of Comet 1888 I.
Astronomical Journal, No. 182, 8:112, October 2.

Filar-Micrometer Observations of Comet 1888 III.
Astronomical Journal, No. 183, 8:119, October 17.

1889   Filar-Micrometer Observations of Comet 1889 I.
Astronomical Journal, No. 187, 8:149, January 22.

PUBLICATIONS OF FRANCIS P. LEAVENWORTH

1889   Proper Motion of Some Double Stars.
Sidereal Messenger, 8:77–80, February.

Observations of the Occultation of Jupiter on March 23, 1889.
Astronomical Journal, No. 191, 8:183, April 15.

Observations of Comet 1888 III.
Astronomical Journal, No. 193, 9:7, May 16.

Observations of Comet e 1888.
Astronomical Journal, No. 196, 9:31, June 19.

Observations of Comet c 1889.
Astronomical Journal, No. 198, 9:46, July 17.

Micrometical Measurements of Double Stars and Other Observations made at the Haverford College Observatory.
Haverford College Studies, 1:19–87.

Observations of Comet e 1888.
Astronomical Journal, No. 205, 9:100, November 25.

1890   Large versus Small Telescopes—A Special Case.
Sidereal Messenger, 9:47–48, January.

Sun-spot Observations.
Astronomical Journal, No. 210, 9:143, February 4.

Observations of Sun-spots and Comets.
Haverford College Studies, 4:17–21.

Observations of Comet d 1889.
Astronomical Journal, No. 210, 9:141, February 4.

Observations of Comet d 1889.
Astronomical Journal, No. 216, 9:192, April 21.

Sun-spot Observations.
Astronomical Journal, No. 222, 10:45, July 21.

1891   Sun-spot Observations.
Astronomical Journal, No. 234, 10:143, January 23.

Personal Error in Observations of Position Angle.
Sidereal Messenger, 10:116–118, March.

Sun-spot Observations.
Astronomical Journal, No. 244, 11:32, July 24.

PUBLICATIONS OF FRANCIS P. LEAVENWORTH

1891   Parallax of Lalande 1196 (South 503).
Haverford College Studies, 10:36–49.

1892   Observations of Double Stars, Comets and Sun-spots in collaboration with Wm. H. Collins and others.
Haverford College Studies, 11:83–108.

Parallax of Delta Herculis.
Haverford College Studies, 11:56–82, January.

Observations of the Periodic Comet of Wolf.
Astronomical Journal, No. 255, 11:120, February 13.

Observations of Double Stars and Comets in collaboration with others.
Haverford College Studies, 12:11–49.

Observations of Comet a 1892.
Astronomical Journal, No. 260, 11:157–158, March 31.

On the Parallax of Delta Herculis from Dembowski's Observations.
Astronomical Journal, No. 262, 11:169–172, April 16.

Observations of Comet a 1892.
Astronomical Journal, No. 264, 11:192, May 3.

On the Proper Motion and Parallax of Delta Equulei.
Haverford College Studies, 12:6–10, June 10.
(Reprinted from Astronomical Journal, No. 270, 12:41–42, July 2)

Observations of Comet a 1892.
Astronomical Journal, No. 269, 12:39–40, June 22.

Measures of Double Stars with 10″ Equatorial of Haverford College.
Astronomical Journal, No. 278, 12:105–110, November 14.

1893   Parallax of O. Argelander 14320.
Haverford College Studies, 12:1–5, February.

Observations of the Parallax of O. Argelander 14320.
Astronomy and Astrophysics, No. 3, 12:206, March.

1894   A Device for Securing a Mercury Surface Undisturbed by Earth Tremors.
Astronomy and Astrophysics, No. 7, 13:597–598, August.

1895   Determination of the Difference of Longitude between the Observatory of the University of Wisconsin and the Observatory of the University of Minnesota by Comstock and Leavenworth.
Printed as a 4-page pamphlet.

1896   Double Star Observations made with the 16″ Refractor of Goodsell Observatory.
Astronomical Journal, No. 382, 16:195–196, August 20.

Micrometical Measure of 70 Ophiuchi.
Astronomical Journal, No. 389, 17:35, December 3.
(Republished in Astronomical Journal, No. 407, p. 178.)

Observations of Comet b 1896.
Astronomical Journal, No. 390, 17:48, December 10.

1897   Observations of Comet g 1896.
Astronomical Journal, No. 401, 17:131, April 27.

Observations of Sappho (80).
Astronomical Journal, No. 406, 17:176, June 18.

Measures of Double Stars.
Astronomical Journal, No. 407, 17:177–180, July 2.

A New Star in the Nebula of Orion.
Astronomical Journal, No. 414, 18:48, October 6.

On the Object in the Nebula of Orion Announced in A. J. 414.
Astronomical Journal, No. 416, 18:62–63, October 27.

1898   Observations of (247) Eukrate, from Photographic Measurements.
Astronomical Journal Nos. 448–450, 19:154, November 22.

1900   Measures of Photographic Plates, Eros and Stars for Solar Parallax.
Conférence Astrophotographique Internationale Circulaire, 12:B141–150, July.

PUBLICATIONS OF FRANCIS P. LEAVENWORTH

1900 Photographic Measures of the Ring Nebula in Lyra and of the Faint Neighboring Stars.
      Monthly Notices, 61:25–29, November.

1903 Parallax of the Sun from Photographs of Eros.
      Astronomical Journal, No. 539, 23:113–115, June 1.

1910 Observations of Comet a 1910.
      Astronomical Journal, No. 613, 26:110, May 20.

1911 Observations of Comet c 1911.
      Astronomical Journal, No. 629, 27:40, November 20.

1912 Double Star Measures.
      Astronomical Journal, Nos. 633–634, 27:71–77, April 20.

      Observations of Comet c 1911.
      Astronomical Journal, Nos. 635–636, 27:97, June 10.

1915 Transit of Mercury, November 7, 1914.
      Astronomical Journal, Nos. 670–672, 28:201, January 22.

      Observations of Comets with 40″ of Yerkes Observatory.
      Astronomical Journal, Nos. 673–674, 29:15, March 15.

      Micrometrical Measures of Double Stars made with the 40″ Refractor of Yerkes Observatory.
      Astronomical Journal, No. 675, 29:17–24, March 22.

      Double Star Measures with the 10½″ Refractor of the University of Minnesota.
      Astronomical Journal, No. 678, 29:41–44, July 12.

1917 Observations of Comets. (With Hugh Wilcox)
      Astronomical Journal, No. 705, 30:77–78, January 8.

      Micrometric Measures of Double Stars made at Yerkes Observatory.
      Astronomical Journal, Nos. 708–709, 30:107–122, February 27, March 20.

1919 Measures of Double Stars with 10½″.
      Astronomical Journal, Nos. 749–750, 32:39–47, March 28, April 18.

Publications of Francis P. Leavenworth

1919   Micrometric Measures of Double Stars made at Yerkes Observatory.
Astronomical Journal, No. 761, 32:129–133, December 17.

1920   Observation of the Eclipse of the Sun, November 22, 1919.
Astronomical Journal, No. 765, 32:165, March 6.

Observations of Comet d 1919.
Astronomical Journal, No. 765, 32:166, March 6.

1921   Observation of the Partial Eclipse of the Sun, November 9, 1920.
(With William O. Beal)
Astronomical Journal, No. 783, 33:130, February 9.

Measures of Double Stars with 10½″.
Astronomical Journal, No. 787, 33:158–162, May 20.

Observations of Comets.
Astronomical Journal, No. 794, 34:12–13, November 28.

1922   Measures of Double Stars with 10½″.
Astronomical Journal, No. 810, 34:151–155, November 24.

1923   Observations of Baade's Comet.
Astronomical Journal, No. 823, 35:59–60, June 26.

1924   Elements of Asteroid Y. O. 18.
Astronomical Journal, Nos. 841–842, 36:15–16, October 15.

1925   Measures of Double Stars with 10½″.
Astronomical Journal, No. 854, 36:105–109, December 14.

1926   Elements and Ephemeris of Asteroid (1045) 1924 TR.
(With Louis Berman)
Astronomical Journal, No. 860, 36:160, May 12.

1927   Measures of Double Stars with 10½″.
Astronomical Journal, No. 881, 37:141–148, June 28.

Made in the USA
Monee, IL
07 July 2026